José Antonio Viana Lopes
Yata Anderson G. Masullo
Tiago Silva Moreira

Interfederative governance in the Metropolis Statute

José Antonio Viana Lopes
Yata Anderson G. Masullo
Tiago Silva Moreira

Interfederative governance in the Metropolis Statute

Study for the new configuration of the Metropolitan Region of Greater São Luís (MA)

ScienciaScripts

Imprint

Any brand names and product names mentioned in this book are subject to trademark, brand or patent protection and are trademarks or registered trademarks of their respective holders. The use of brand names, product names, common names, trade names, product descriptions etc. even without a particular marking in this work is in no way to be construed to mean that such names may be regarded as unrestricted in respect of trademark and brand protection legislation and could thus be used by anyone.

Cover image: www.ingimage.com

This book is a translation from the original published under ISBN 978-613-9-62708-0.

Publisher:
Sciencia Scripts
is a trademark of
Dodo Books Indian Ocean Ltd. and OmniScriptum S.R.L publishing group

120 High Road, East Finchley, London, N2 9ED, United Kingdom
Str. Armeneasca 28/1, office 1, Chisinau MD-2012, Republic of Moldova, Europe
Printed at: see last page
ISBN: 978-620-7-74917-1

SUMMARY

PRESENTATION

At a time when the state of Maranhao is going through a new period of state administration, it is necessary to analyze local and regional complexities and singularities in a holistic way in order to meet the growing social demand in all aspects of our reality. To this end, expanding our knowledge of the territory is *a sine qua non* condition for developing public policies that are capable of dealing with the intense and dynamic process of urbanization underway in our cities.

In this scenario, we find the Metropolitan Region of Greater Sao Luis, where social, economic and environmental data indicate the permanence and increase of the concentration of services and investments, requiring the reorientation of urban and institutional dynamics towards the decentralization of resources and local public policies as a way of avoiding the fragmentation of the socio-political-spatial fabric in the metropolis.

In this way, it becomes imperative that the public interest prevails over individual interests, reorienting public policies towards continuous planning processes and the decentralization of services and investments as conditions for solving problems of common interest, improving the quality of life of the population.

By providing information and enabling basic analysis for planning, the Secretary of State for Cities and Urban Development - SECID, through the publication of this Technical Study for the New Configuration of Greater Sao Luis, is taking another important step towards overcoming the barriers and obstacles imposed on urban policies and inter-federative management of the Greater Sao Luis Metropolitan Region.

Flâvia Alexandrina Coelho Almeida Moreira

SECRETARY OF STATE FOR CITIES AND URBAN DEVELOPMENT

INTRODUCTION

The intense and rapid process of urbanization in Brazil in recent decades poses a major challenge for the management of large cities, as it is a dynamic and complex process resulting from social, economic and political interaction intrinsically linked to the diverse interests of our society.

On the threshold of the 1970s in Brazil, there was an attempt at urban restructuring through a centralizing wave of the then military government, with the aim of planning the large urban centers in a hierarchical way without there being a broad discussion within the agglomerated areas to assess which sectors could be developed in an integral way or in a strategic plan that could balance local inequalities. From this perspective, polarized regions were created, called Metropolitan Regions, following the trend in developed countries with the aim of solving problems common to conurbed cities.

Federal regulation began with the 1967 Constitution, but due to political disputes it was consolidated in 1973 with Federal Complementary Law 14, which institutionalized the metropolitan regions of Sao Paulo, Belo Horizonte, Porto Alegre, Recife, Salvador, Curitiba, Belém and Fortaleza. According to Rolnik; Somekh (2000), at that time, the establishment of metropolitan regions took place from the perspective of the authoritarian regime, choosing the city as a problem and obeying the economic development strategy adopted by the military government, with a reduction in the autonomy of states and municipalities.

At the same time, the historical process of the formation of Brazilian cities has been discontinuous, but centered on the country's economic cycles. In this way, to understand metropolization, particularly in Brazil today, is to understand a little more about the workings and dynamics of contemporary capitalism, particularly from the point of view of how urban social space is produced in it (IPEA, 2010).

The challenge facing the configuration and implementation of metropolises, especially in Brazil, is to plan in a non-rationalist and flexible way, understanding that history is a complex mixture of determination and indeterminacy, of rules and contingencies, of structural conditioning and degrees of freedom for individual action, in which the expected is often sabotaged by the unexpected - which makes any planning both necessary and risky (SOUZA, 2010).

What has become a central issue in recent studies on Metropolitan Regions is the way in which the actions of the agents involved in the planning and management of the MR's are treated and analyzed, no longer thinking of the place in a segregated and isolated way, but broadening and making more flexible the interests between the municipalities and the area of influence. This problem becomes even more far-reaching when we consider that Brazil has 71 Metropolitan Regions (MRs), 4 Urban Agglomerations (UAs) and 3 Integrated Development Regions (RIDEs) (OBSERVATÓRIO DA METRÓPOLE, 2015).

Brazil's metropolitan regions are home to approximately 110 million people, 60% of the country's GDP, in a space equivalent to 2% of the country's territory (OBSERVATÓRIO DA METRÓPOLE, 2015). Among them is the Metropolitan Region of Greater Sao Luis - RMGSL, which has a population of 1,605.305 inhabitants, concentrating 39.4% of the GDP of the state of Maranhao (IMESC, 2016; IBGE, 2016), and still has structural problems, such as income inequality - with approximately 9% of the population living below the extreme poverty line - a high housing deficit, more than 20% of the population living in subnormal agglomerations, increasing rates of violence, urban mobility problems, vagueness of municipal boundaries, as well as problems and gaps in its creation law and in the process of institutionalization and implementation.

This study was carried out by technicians from the Deputy Secretariat for Metropolitan Affairs of the Maranhao State Secretariat for Cities and Urban Development, with the aim of supporting and substantiating public discussions on legal and institutional changes capable of making the Greater Sao Luis Metropolitan Region effective, in the light of the Metropolis Statute (Law 13.089 of January 2015)[1] . The aim was therefore to gather data, analyze and characterize Greater Sao Luis socially, environmentally and economically, based on the territorial transformations resulting from urban planning and the proposed implementation of shared metropolitan management.

It should be emphasized that the analysis of social, political, economic and environmental indicators in this study marks the beginning of a collective reflection on the barriers and obstacles historically placed in the way of urban policies and the intergovernmental management of a metropolitan region. In this way, the role of the state is highlighted as an element of integration and the dynamic engine of the process which, based on the

---

1 The Study was prepared in 2015 and updated in December 2017 for publication, adding the approved legislation.

overlapping layers of information, organizes the planning system to promote regional and metropolitan development.

5

TECHNICAL ANALYSIS AND PROCEDURES

Delimitation

The RMGSL is a complex territory characterized by diversity in social, economic and environmental issues. The current delimitation of the RMGSL is based not only on criteria of urban homogeneity, but also on territorial complementarity and cohesion, as well as the need for economic integration and the reduction of social inequalities. In this way, the new metropolitan configuration of Greater São Luis is technically supported by the following factors:

1.   Economic concentration based on the diversification of industrial activities and services, with major investments and companies such as ALUMAR, VALE, the Port of Itaqui, and the planned installation of Petrochemicals and Steel, among others.

2.   Tourism linked to the existence of cultural and natural heritage, inserted in a coastal region and made up of 7 conservation units, covering 63.5% of its territory, totaling 3,973.87 km^2 , as well as municipal and private protected areas. It is worth highlighting the recognized cultural heritage linked to its urban and architectural ensembles, remnants of the 17th, 18th and 19th centuries, and various traditional manifestations.

Criteria and indicators

The composition of the RMGSL with 13 municipalities has technical and strategic articulation criteria. The addition of new municipalities to the Greater São Luis Metropolitan Region must be based on prior technical studies, to be drawn up by a public research institution with notorious knowledge and experience in regional and urban studies, and must be approved by the Metropolitan Collegiate, for subsequent submission to the Maranhão Legislative Assembly, taking into account the following criteria:

I - functional articulation between municipalities, with contiguity and/or discontinuity of the area of occupation (ports, airports, complex services, dormitory cities, research and innovation, major economic and infrastructure investments, landfills, water sources, etc.);

II   - insertion in the region of influence of the municipality of Sao Luis, according to the Brazilian Institute of Geography and Statistics - IBGE (REGIC);

III   - annual population growth rate above the state average (1.52% p.a. between 2000 and

2010);

IV   - existence or need for public functions of common interest;

V   - high tourist interest, environmental protection and cultural valorization;

VI   - significant commuting of the population to work and/or study.

The implementation and expansion of the Greater Sao Luis Metropolitan Region meets the aforementioned criteria with regard to its economic and demographic dynamics, highly diverse urban and regional functions, specialization, economic integration and the reduction of social inequalities.

The list of selected indicators demonstrates the need for economic and social decentralization of the RMGSL based on its industrial, commercial, services, natural resources and historical and cultural heritage.

The intense process of conurbation in the polarizing municipalities of the RMGSL along the MA 201 (Sao Luis/Sao José de Ribamar); MA 202 (Sao Luis/Paço do Lumiar), MA 203 (Sao Luis/Raposa); and MA 204 (Raposa/Paço do Lumiar/Sao José de Ribamar) requires institutional coordination to solve problems linked to public functions of common interest.

It is also worth noting the integration of the municipalities which are under the direct influence of the current unequal and concentrating economic system and are interconnected through the functional relationship provided by the Praia Grande Ferry Terminal and Port Terminal (Sao Luis/Alcântara); BR 135 (Sao Luis/Bacabeira/Santa Rita); MA 402 (Bacabeira/Rosàrio/Axixà/Morros/Icatu); MA 020 (Morros/Cachoeira Grande/Presidente Juscelino).

Its main features are:

■   Strategic location connecting the metropolitan region to major importing centers for Brazilian products such as Europe and the United States, which saves on fuel and reduces the delivery time for goods coming from Brazil through the Port of Itaqui, which is the second deepest in the world and one of the busiest for foreign trade in Brazil.

■   It houses the Alcântara Launch Center - CLA, which is considered one of the best in the world due to its geographical location, precisely because of its proximity to the equator, which favors the thrust of the launchers and the fuel economy of the propellant used in the rockets. This factor is added to the climatic conditions of the place, which has a stable climate, a well-defined rainfall regime and winds within acceptable limits, making it possible to launch rockets in practically every month of the year.

■ The RMGSL is connected to the interior of the state by the Carajás Railway and also to the neighboring states of Parà, Tocantins and Piaui, which makes it possible to interconnect and transport the production of ore and agriculture from the interior of the country to the Port of Itaqui. By road, the central axis of the region is the BR-135 highway linking it to the mainland, and it also has the Cunha Machado airport, the largest and busiest airport near the Lençôis Maranhenses National Park.

■ Economic importance at the state level, with an estimated Gross Domestic Product (GDP) of R$30,252,442.40 billion in 2014, equivalent to 39.4% of the product generated in the state. Of this, 88% is concentrated in Maranhão's capital.

■ Population of 1,605,305 inhabitants in 2016 - ranks among the 20 largest institutionalized Metropolitan Regions in Brazil.

HISTORY OF GREATER SAO LUIS

Between the 1970s and 1990s, there was a rapid process of urbanization in the capital of Maranhao, due to factors that attracted a large population to the municipality of Sao Luis, such as the territorial policy linked to the Grande Carajás Program. This was developed with the installation of Vale do Rio Doce and ALUMAR, with investments of around R$ 224 million, which transformed the region into a labor attraction hub, attracting contractors and other public and private investments, which aggravated urban problems.

The region went through an intense process of peripheralization during this period and the spatial cut-outs followed the logic of occupation of public lands or unoccupied plots, giving rise to the process of disordered occupation in Sao Luis:

These projects with a developmentalist discourse have also caused a large population contingent from the interior of the state, neighboring states and other regions to move to the capital of Maranhão, thus inducing an expansion of the service sector. The industrial complexes that set up in Maranhao, despite their large investment, do not provide the same number of direct jobs in the state and in its capital Sao Luis (MOREIRA, 2013, p. 40).

In the course of this process, several attempts were made to raise discussion about the management of a city that is growing at an accelerated pace compared to the other municipalities on the island, but the measures adopted failed from a practical point of view.

In 1987, the Fòrum de Debates sobre a Grande Sao Luis, organized by the Secretariat of Work and Urban Development, proposed the *Carta de Urbanismo da Grande Sao Luis,* a document that included "metropolization" as one of the themes of society's demands. This demand was adopted by the State Constitution two years later.

Created in the 1989 State Constitution, it wasn't until the end of the 1990s that a Complementary Bill was drawn up to define the scope, organization and functions of the RMGSL. This project was approved in 1998, through State Complementary Law No. 038/98, encompassing the municipalities of Ilha do Maranhao (Sao Luis, Sao José de Ribamar, Paço do Lumiar and Raposa), authored by Deputy Francisco Martins. As well as establishing the RMGSL, this provision created the Council for the Administration and Development of the Metropolitan Region of Greater Sao Luis (COADEGS).

In 2002, the Sao Luis City Council created the Municipal Secretariat for Articulation and

Metropolitan Development (SADEM)[2] , which to date is the only municipal body with powers relating to metropolitan management. In 2013, the Sao Luis City Council set up the Metropolitan Affairs Commission[3] as a permanent commission that has been active throughout the process of discussing the metropolitan region.

Five years passed without the installation of the Council provided for in LCE No. 038/98. In 2003, Complementary Law No. 69/2003 was passed, authored by Deputy Alberto Franco, which gave the metropolitan region a new structure by adding the municipality of Alcântara, as well as reconfiguring the composition of COADEGS, and proposing the creation of a state autarchy and the region's Development Fund, which were vetoed by the state executive branch[4] .

At this time, discussions about the formation of Greater Sao Luis were resumed, but, centered on the debate about the definition of municipal boundaries (mainly between Sao Luis and Sao José de Ribamar), they ended up making no progress. Whether due to a lack of political will on the part of the state government, a lack of knowledge about the structures and practices of metropolitan management, or even a lack of clear references in national urban policy for metropolises, the argument was created that the RMGSL would be unfeasible as long as the problems and uncertainties between municipal boundaries persisted.

Several attempts have been made to tackle the issue of municipal boundaries, since the events coordinated by the Brazilian Institute of Geography and Statistics (IBGE), with the participation of the municipalities (especially SADEM) and the State Legislative Assembly (ALEMA) through its Municipal Affairs and Regional Development Commission in 2005.

At that time, the debate ended up focusing on the contexts of municipalities which, guided by Law No. 10.257/2001, the Statute of Cities, and by the actions of the Ministry of Cities, were trying to implement participatory planning processes for revising or drawing up their master plans. It was only in 2010 that the operational boundary line between the municipalities on the island of Maranhao was defined, supported by the Maranhense Institute of Socio-Economic and Cartographic Studies (IMESC), which has been used to date and taken as the basis for a new attempt to legally establish inter-municipal boundaries

---

2 Through Law No. 4128 of December 23, 2002, sanctioned by Mayor Tadeu Palacio.
3 Created by Amendment No. 5/2013 to Resolution No. 337/1983.
4 MARANHAO. Message No. 136/2003. Partial veto of State Complementary Law No. 69/2003. Diario oficial, n. 248, 22 dec. 2003.

in Greater Sao Luis, which is currently underway.

In 2007, the Father Marcos Passerini Defence Centre (CDMP) took up the role of civil society on the issue and organized the seminar *Metropolitan Region of Greater São Luis - Impasses and Implications for Public Policies*, discussing aspects of the process of metropolization and the development of *public policies* for the region.

Between 2008 and 2010, some institutions of organized civil society, articulated by the Maranhao State Engineers' Union (SENGE), supported by the National Federation of Engineers (FNE) and the Maranhao State University (UEMA), resumed and strengthened the discussions on new bases. To this end, they launched the *Metropolitan Charter of Greater São Luis* (2008), proposing and establishing the Metropolitan Forum of Greater São Luis. The Forum held two major meetings, in November 2008 in Sao José de Ribamar, and the following November in Paço do Lumiar.

This initiative brought together public bodies directly related to the issue, such as SADEM representing the Sao Luis City Hall, managers and representatives from the municipalities of Sao José de Ribamar and Paço do Lumiar. However, once again the lack of political coordination between the municipalities (particularly the hub city of Sao Luis) and the state government limited the movement's results. The many demands in different areas (sanitation, solid waste, mobility and public safety) also diluted efforts and took the focus away from the necessary management structures that needed to be created.

In any case, the state government needed to respond to the issue, which was now taking on the dimension of a social and political demand. This led to the creation of the Deputy Secretariat for Metropolitan Affairs (SAAM), through Decree No. 27.209, of January 3, 2011, linked to the Civil House. At this time, SAAM added to its team some of the professionals linked to SENGE's board who had accumulated experience from the discussions at the Greater Sao Luis Metropolitan Forum.

However, despite organizing the creation of a State Metropolitan Policy Committee to define the *Guidelines* for the *Implementation of Metropolitan Management, with* the collaboration of the Paraná Institute for Economic and Social Development (IPARDES)[5] , The Deputy Secretariat did not have the executive functions of a metropolitan agency, nor did it have the structure and resources capable of proposing and executing Public Functions

---

5 Through the collaboration of Prof. Dr. Rosa Moura.

of Common Interest, so it began to act mainly in an attempt to articulate municipalities and produce technical knowledge about the reality of the region[6] .

With this format, the state government has shown no interest in implementing other bodies linked to the issue, cooling the dynamics of the debate and making it centralized and punctual.

The Seminar *São Luis + 400 Years: Discussing the Central Region and the Metropolis in the Light of the City Statute*, held in August 2012 with the participation of the State Council of Cities (CONCIDADES) and representatives of public authorities and civil society organizations, in commemoration of the city's anniversary, resulted in the *Upaon-Açu Charter* which, among other issues, calls for the creation of management and social control instruments for the metropolitan region, pressuring the state to take action again (BRASIL, 2014).

In response and in the wake of expectations of the installation of major projects of national interest in the region, a new change in the configuration of the RMGSL was proposed, based on the formulation of State Complementary Law No. 153 of April 10, 2013, authored by Congressman Jota Pinto, which incorporated the municipalities of Bacabeira, Rosàrio and Santa Rita. This change was implemented in view of the possible installation of Petrobras' Premium I refinery in the municipality of Bacabeira and its impacts on the region.

This change in legislation was accompanied by a technical report issued by SAAM and sent to the Legislative Assembly by means of letter 377/2013. A new request to change the configuration of the RMGSL was submitted shortly afterwards by means of letter 214/2013.

However, contrary to what happened previously, State Complementary Law No. 161, of December 3, 2013, authored by Congressman Eduardo Braide, was passed and approved without SAAM's monitoring or technical analysis, incorporating the municipality of Icatu into the RMGSL.

These events demonstrate the limitations of the body responsible for establishing and articulating metropolitan governance in the state, where the political context overrides social

---

6 Such as the *Guidelines for the Implementation of Metropolitan Management* (SAAM, 2012), the research *Characterization and Comparative Analysis of Metropolitan Governance in Brazil: institutional arrangements for metropolitan management (Component 1): Metropolitan Region of Greater São Luis*, published by IPEA (BRASIL: 2014) and the report *Results of the HDI-M of the Human Development Units of the Metropolitan Region of São Luis*, also published by IPEA (IMESC: 2015).

demands, even in the context of changes in the territorial dynamics of the region, increasing territorial and socio-economic discontinuities.

In this way, the configuration of the current RMGSL reflected the weaknesses and gaps in the legislation approved between 1998 and 2013[7] . These gaps include the unconstitutionality of provisions requiring the approval of municipal councils for the inclusion of municipalities in the region (such as the State Constitution itself and LCEs No. 038/1998 and 069/2003)[8] , the lack of technical criteria for including municipalities in the RM, the lack of planning instruments for public functions of common interest, the creation but not implementation of a Development Council with deliberative functions without, however, providing for an executive body, a common fund for allocating resources, or even popular participation in the Council, which was never installed.

In general, it can be seen that the legislation, focused on the inclusion of new municipalities, did **not make it possible to set up the institutional arrangements needed to manage the metropolitan region**. What stands out in this context of interfederative governance in Maranhao is the social mobilization around metropolitan issues, the lack of dialogue and political articulation between the state government and the municipalities (particularly Sao Luis) and the resulting limitation on the actions of the (few) metropolitan management bodies created (SADEM and SAAM). Therefore, a situation has been maintained in which the region suffers the impacts generated by the implementation of services and undertakings that affect everyone, without integrated policies or actions to solve these problems. From this socio-political context, in 2015 the state government took on the challenge of making the metropolitan region's shared management arrangement work.

With the approval of Law No. 13,089 of January 12, 2015, the Metropolis Statute, the National Urban Development System (SNDU) established the criteria and guidelines that the federal government adopts for the development policy of metropolitan regions. With the entry into force of the Statute, there was inevitable disquiet about the adjustments and adequacy of Brazil's metropolitan regions, since only those territorial units that implemented their management plans would be recognized, according to Article 8:

---

7 State Complementary Law No. 038 of January 12, 1998; State Complementary Law No. 069 of December 23, 2003; State Complementary Law No. 153 of April 10, 2013; and State Complementary Law No. 161 of December 3, 2013.
8 According to the Federal Supreme Court (STF), this provision goes against the provisions of the 1988 Federal Constitution in its Article 25, paragraph 3, which only requires the approval of a State Complementary Law to set up a Metropolitan Region.

Art. 8 Inter-federative governance of metropolitan regions and urban agglomerations will comprise its basic structure:

I - an executive body made up of representatives of the executive branch of the federal entities that make up the urban territorial units;

II - a deliberative collegiate body with civil society representation; III - a public organization with technical advisory functions; and

IV - integrated resource allocation and accountability system (BRASIL, 2015. P. 3).

In this scenario, we have at the same time the inauguration of a new form of territorial management, now regulated and adopted on a national scale, and on the other hand, the old institutional dilemmas of metropolitan management such as the political fragmentation that since the creation of Greater São Luis (1998) has hindered the effectiveness of the metropolization process and instituted the paradox between functionality and institutionality (MASULLO; LOPES, 2017).

The Statute of the Metropolis establishes the basic structure for the interfederative governance of the Metropolitan Regions, providing for an executive body, a collegiate body with civil society representation, a public organization with technical-consultative functions and an integrated system for allocating resources and rendering accounts. This is the legal framework that guided the institutional restructuring of Greater Sao Luis.

In one of the first administrative actions in 2015, the state government decided to keep SAAM in its structure, as the body responsible for coordinating and implementing interfederative governance in the state's metropolitan regions, following Decree No. 28,884 of February 20, 2013, which transferred it from the Civil House to be permanently installed in the State Secretariat for Cities and Urban Development - SECID. With the change of institutional link to SECID and a new strategic vision from the current state government, a restructuring of the RMGSL was consolidated through meetings between representatives of the state government, municipalities and civil society.

Thus, shortly after the approval of the Metropolis Statute, on February 4, 2015, the governor held the first meeting with the mayors of the metropolitan region since its creation in 1998, to set up the **Greater São Luis Metropolitan Region Working Group** (Annex 1), to revise Law No. 69/2003 (Figure 01).

To support the discussions and definitions of the Working Group, SECID's multidisciplinary technical team prepared the **Technical Study for the New Configuration of Greater São Luis**, gathering available data, cadastral information and cartography on various aspects (economic, social, cultural, environmental) of **the** local reality, with the aim of putting together a current panorama of the metropolitan region.

Figure 01 - Greater Sao Luis Working Group.

Source: SAAM/SECID collection, 2015.

Fulfilling the important role of articulating public interests, political conditions and projects for regional development, the Government of the State of Maranhao, through Message No. 056/2015, forwarded to the Legislative Assembly Complementary Bill No. 004/2015, the result of the work of teams that brought together more than 80 (eighty) people, including mayors, councillors, advisors and technicians from all the municipalities, as well as state institutions such as the State University of Maranhao, the Maranhão Institute of Socioeconomic and Cartographic Studies (IMESC) and the Federation of Municipalities of the State of Maranhao (FAMEM), under the coordination of the State Secretariat for Cities and Urban Development (SECID).

This process resulted in the drafting and approval of **State Complementary Law No. 174 of May 25, 2015** (Annex 2), which provides for the establishment and management of the Metropolitan Region of Greater Sao Luis and repeals State Complementary Laws No. 038 of January 12, 1998, No. 069 of December 23, 2003, No. 153 of April 10, 2013, No. 161 of December 3, 2013 and other provisions to the contrary or that are incompatible.

LCE No. 174/2015 provides the Metropolitan Region of Greater Sao Luis with the management and planning tools necessary for its effective operation, reconfiguring the region and adapting our legal framework to the recommendations and guidelines of the Metropolis Statute.

To reinforce and broaden this debate, in 2016 SECID organized a series of events open to the community called *Metropolitan Dialogues*, with the aim of creating a dynamic of debate and discussion of ideas in search of effective solutions to the problems faced by the municipalities of the RMGSL (Figure 02).

The first event dealt with the theme of "Sustainable Urban Mobility: the role of public space in the restructuring of cities", led by Pernambuco architect and urban planner Milton Botler. A second event dealt with the theme of "Revising Sao Luis' Urban Legislation" and was led by the team from the City Institute (INCID) of Sao Luis City Hall. The third event in this series dealt with the theme "Metropolitan Strategy for Solid Waste Disposal", a proposal presented by the SECID team.

Figure 02 - *Metropolitan Dialogues:* Sustainable Urban Mobility.

Source: SAAM/SECID collection, 2016.

Based on the approved legislation, SECID sent requests for standardization and reduction of telephone tariffs (to ANATEL), transport (to the State Transport Agency - MOB) and expansion of housing finance (to CAIXA) in the Greater Sao Luis Metropolitan Region.

In order to align the positions of different institutions in their approach to the RMGSL,

SECID's technical team and managers took part in commissions and working groups on the regionalization of the state and the redefinition of municipal boundaries:

a. Working Group on the **New Regionalization of the State of Maranhao**, of the Instituto Maranhense de Estudos Socioeconômicos e Cartograficos - IMESC.

b. Working Group on **Municipal Boundaries on the Island of Maranhao,** of the Municipal Affairs and Regional Development Commission of the Legislative Assembly of Maranhao.

c. **State Commission for Ecological-Economic Zoning of the State of Maranhao**, in accordance with Decree No. 29.359, of September 11, 2013 and Law No. 10.316, of September 17, 2015.

Following the approval of LCE 174/2015 and the discussions held at the Participatory Metropolitan Dialogues and Territorial Listening events of the Participatory PPA, SECID proposed and included in the state's 2016-2019 Multiannual Plan the <u>Metropolitan Regions Implementation and Restructuring Program,</u> with the aim of implementing interfederative governance instruments, sectoral plans, strategic projects and works for the effectiveness and development of the state's metropolitan regions. The first action of this program, called *Interfederative Governance,* includes setting up the bodies needed to draw up and approve the RMGSL's Integrated Development Master Plan.

In a parallel effort to the discussions on the review of legislation, the problems and solutions of the region's public functions of common interest and the inclusion of the issue of metropolitan regions in the state's PPA, SECID's technical team also dedicated itself to drawing up the basic documents for regulating the instruments and contracting the plans provided for in LCE No. 174/2015:

a. Preparation of a draft bill to regulate the **Metropolitan Executive Agency**;

b. Preparation of the draft State Decree regulating the **Metropolitan Development Fund**;

c. Preparation of **a Study on Municipal Contribution Capacity to the RMGSL Fund**, in partnership with the Instituto Maranhense de Estudos Socioeconômicos e Cartograficos - IMESC.

LCE No. 174/2015 assumes the PDUI under the name of Integrated Development Master

Plan - PDDI (Art. 24, paragraph 1), and determines its precedence over other sectoral plans in the region:

Art. 24: The Integrated Development Master Plan - PDDI will contain the guidelines for metropolitan planning, including for metropolitan sector plans and local sector plans [...] (LCE n°174/2015, Art. 24).

Adding that:

Once the procedures set out in the preceding paragraph have been taken into account, the Metropolitan Executive Agency of Greater Sao Luis will issue a decision:

I   - the metropolitan sectoral plan for housing and land regularization;

II   - the metropolitan sectoral plan for urban mobility;

III   - the metropolitan sector plan for basic sanitation (water supply, sewage, drainage and solid waste);

IV   - other metropolitan sectoral plans, relating to public functions of common interest, under the terms of a decision by the Metropolitan Collegiate (LCE n°174/2015, Art. 24, Paragraph 3).

Therefore, following the legal precepts, the Secretary of State for Cities and Urban Development (SECID) drew up the Terms of Reference **for** contracting **the Integrated Development Plan (PDDI) for Greater São Luis, as** well as a **draft technical cooperation agreement** for drawing up the PDDI with the State University of Maranhao (UEMA) and the Maranhão Institute of Socioeconomic and Cartographic Studies (IMESC).

In addition, the **Technical Study for the New Configuration of Greater São Luis,** prepared by SECID's technical team, is an important reference document for the Diagnosis to be carried out in the PDDI. Between 2015 and 2016, SECID also prepared studies and reference documents for the RMGSL's main sectoral plans. These are:

a. Preparation of a Technical Study: **Solid Waste Management in Greater São Luis**, a reference document for the proposal of a "Metropolitan Solid Waste Disposal Strategy";

b. Preparation of the Terms of Reference for the contracting of the **Origin-Destination Study of** the Metropolitan Mobility Plan;

c. Preparation of the Terms of Reference for the contracting of the **Metropolitan Solid Waste Management Plan.**

With these actions, the state government is respecting and implementing the provisions of the Metropolis Statute, which stipulates the implementation of an inter-federative management structure at different levels as a necessary condition for drawing up and approving the plan:

§ Paragraph 4. The plan provided for in the caput of this article shall be drawn up within the inter-federative governance structure and approved by the deliberative collegiate body referred to in item II of the caput of Article 8 of this Law, before being sent to the respective state legislative assembly (Law 13.089/2015, Article 10, paragraph 4).

Reinforcing this position expressed in federal law, LCE 174/2015 indicates the Metropolitan Collegiate as the body responsible for its preparation, monitoring and approval, including its competencies:

I - promote the preparation, monitoring and approval of the Integrated Development Master Plan and Sector Plans, as well as ratifying any revisions that may be necessary (LCE No. 174/2015, Art. 8, Subsection I).

Therefore, in an effort to comply with the recommendations of Law 13.089/2015, and seeing the opportunities of a new political conjuncture with the recent election of municipal mayors, the state government held a first meeting on the "Metropolitan Region of Greater Sao Luis" with the elected mayors of the municipalities of Maranhao Island on October 7, 2016.

Through Provisional Measure No. 229 of February 2, 2017, the State Government changed the organizational structure of the public administration of the Executive Branch of the State of Maranhao, creating the Metropolitan Executive Agency linked to the Civil House, with the attributions set out in art. 15 of LCE 174/2015 and in accordance with the guidelines and determinations of Law No. 13.089/2015, the Metropolis Statute (See Table I).

[a]As a direct result and necessary offshoot of these initiatives, the 1st Metropolitan Collegiate Meeting was held on February 8th, followed on the same day by the 1st Metropolitan Governance Workshop: Metropolitan Experiences, an event promoted by SECID that brought together managers, technicians, academics and civil society to discuss issues related to metropolitan planning (PDDI), mobility and solid waste management (Figure 03).

Figure 03 - la Meeting of the Metropolitan Collegiate.

Source: SAAM/SECID collection, 2006.

The first workshop included a presentation by the Belo Horizonte Metropolitan Development Agency (AGEM-BH) on the process of drawing up the PDDI for Greater Belo Horizonte, with a focus on raising local managers' awareness of the issue, since the Metropolitan Collegiate is already discussing the authorization to draw up the PDDI and the call for the Metropolitan Conference.

Table I - Legislation relating to the Glande Sà Luis Metropolitan Region (1998 - 2015).

LAW	SUMMARY	PROJECT	MESSAGE	PUBLICATION	AUTHOR	GOVERNMENT
Law State Complementary Law No. 038, of January 12, 1998.	Provides for the Greater Sào Luis Metropolitan Region.	State Supplementary Bill No. 04/95		D O Eno 115 of 22/01/1998	Francisco Martins MP	Roseana Samey
State Complementary Law No. 069, of December 23, 2003.	Gives new wording to Complementary Law No. 038. of January 12, which provides for the Metropolitan Region of Greater São Luis and makes other provisions	State Supplementary Bill No. 09/03.	MESSAGE No. 136/2003, OF DECEMBER 18, 2003, PARTIALLY VETOING ARTICLES. 7, 10 AND PARAGRAPHS, 11 AND SOLE PARAGRAPH, 14, 15. 17, 18 AND SOLE PARAGRAPH, 19 AND 21	DOE No. 253 of 30/12/2003	Deputy Alberto Franco	José Remaldo Tavares
Law	Amends Complementary Law	State	MESSAGE No. 23/2013.	DOE of no070 of	Deputy Jota	Roseana Samey

State Complementary Law No. 153, of April 10, 2013.	No. 069, of December 23, 2003, which deals with the Greater São Luis Metropolitan Region.	Supplementary Bill No. 10/11.	OF APRIL 10, 2003 PARTIALLY VETOING ARTICLES 1 AND TOTALLY VETOING ARTICLES 2 AND 3	11 '04/2013	Pinto	
State Complementary Law No. 161 of December 3, 2013.	Gives new wording to article 1 of Complementary Law no. 38. of January 12, 1998, which provides for the Metropolitan Region of Greater São Luis and makes other provisions (to include the municipality of Icatu)	State Supplementary Bill No. 07/13		DOE No. 242 of 12/12/2012	Deputy Eduardo Braide	Roseana Samey
State Complementary Law No. 174, of May 25, 2015.	Provides for the establishment and management of the Greater São Luis Metropolitan Region and repeals State Complementary Laws No. 038 of January 12, 1998, No. 069 of December 23, 2003, No. 153 of April 10, 2013, No. 161 of December 3, 2013 and other provisions in centrano	State Supplementary Bill No. 004/15	MESSAGE NO. 056 OF APRIL 27, 2015.	DOE No. 064 of 28/04/2015	Executive Branch	Flavio Dmo

Source: SAAM SECID. 2015.

Progress and discussions regarding the restructuring of the RMGSL are being monitored through periodic presentations by the Assistant Secretary for Metropolitan Affairs to the Maranhao State Council of Cities - CONCIDADES, in ordinary and extraordinary meetings:

a.   **25ª Meeting of the State Council of Cities - CONCIDADES-MA**. Presentation of the work of the RMGSL WG. Caxias, June 10-12, 2015.

b.   **28ª CONCIDADES Ordinary Meeting**, held on April 18, 19 and 20, 2016, in Cururupu (MA);

c.   **29ª CONCIDADES Ordinary Meeting**, held on July 27, 28 and 29, 2016, in Sao Luis (MA);

d.   **30th Ordinary Meeting of CONCIDADES**, held on October 24, 25 and 26, 2016, in Sao Luis (MA).

e.   **31st Ordinary Meeting of CONCIDADES**, held on December 13 and 14, 2016, in Sao Luis (MA).

The institutionalization and effective implementation of inter-federative governance in the Metropolitan Region of Greater Sao Luis will guarantee the support of the Union, with the contribution of federal resources for sectors and projects of regional interest.

With the implementation of the Collegiate and the creation of the Metropolitan Agency, the Participatory Council, a Development Fund and the preparation of an Integrated Development Plan for the region, the municipalities will have the necessary instruments to tackle common problems that today cannot be solved without coordination and collaboration with other municipalities.

THE NEW METROPOLITAN CONFIGURATION

The new configuration of the Greater Sao Luis Metropolitan Region proposed in LCE No. 174/2015 (Figure 04) incorporated four more municipalities with two clear objectives: to consolidate the current configuration by providing territorial continuity as the Statute of the Metropolis proclaims, through the integration of the municipalities of the Lower Munim region (Axixà, Cachoeira Grande, Morros and Presidente Juscelino), and to enable their development process with a focus on specific projects in the areas and services of common interest between the municipal entities, based on the strategic vision of the State Government, acting to decentralize resources and reduce social inequalities.

Figure 04: Location map of the Greater São Luís Metropolitan Region

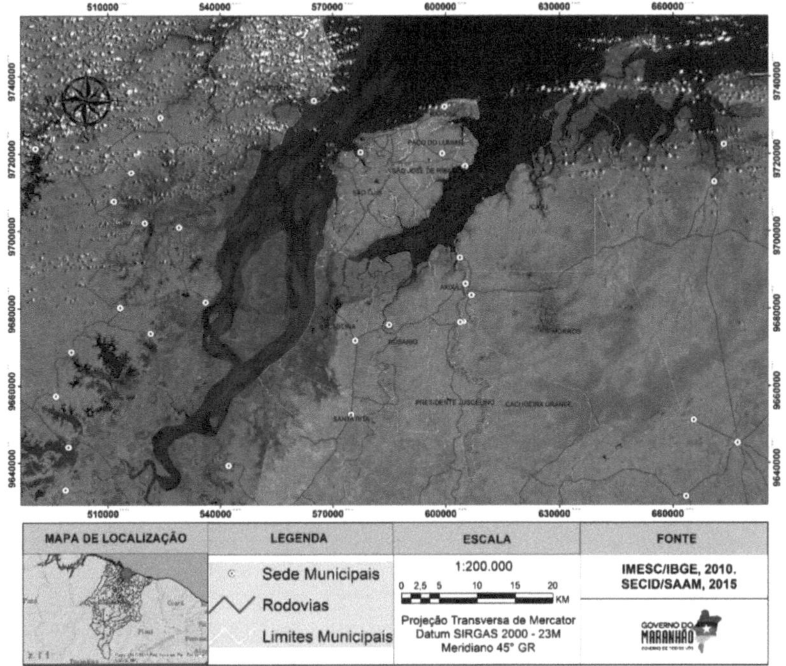

MAPA DE LOCALIZAÇÃO	LEGENDA	ESCALA	FONTE
	Sede Municipais	1:200.000 0 2,5 5 10 15 20 KM	IMESC/IBGE, 2010. SECID/SAAM, 2015
	Rodovias	Projeção Transversa de Mercator Datum SIRGAS 2000 - 23M Meridiano 45° GR	GOVERNO DO MARANHÃO
	Limites Municipais		

Source: SECID/SAAM, 2015.

The map above shows the new configuration of the RMGSL, which will be made possible and effective beyond pre-defined structural concepts with a volume of specific projects and investments for the region, given the extreme need to structure municipalities such as Presidente Juscelino, Cachoeira Grande, Morros, Axixà and Icatu, which are still undergoing an incomplete urbanization process and have poor public services, revealing the great spatial and urban discontinuity of the RMGSL.

Figure 05: Map of the RMGSL Region of Influence

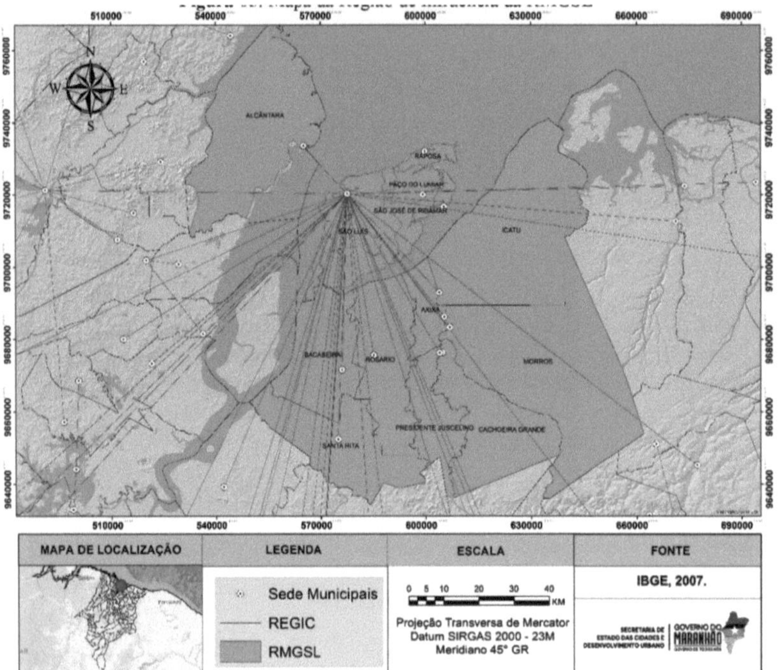

Source: SAAM, 2016.

It should be noted that the new configuration of the RMGSL still has a fragmented structure and major challenges beyond its institutional implementation and pre-defined structural concepts. It is therefore necessary to make actions feasible and to implement specific projects and investments for the region, given the extreme need to structure municipalities such as Presidente Juscelino, Cachoeira Grande, Morros, Axixà and Icatu, which are still undergoing an incomplete process of urbanization and have poor public services, focused on the primary sector with low socio-demographic and economic dynamics, revealing the great spatial and urban discontinuity of the region.

In this sense, the state government, together with the municipal players, must face up to the challenge of making the RMGSL a reality through planning and management tools that will integrate and consolidate the metropolitan region, with mechanisms and infrastructure geared towards this new configuration. The aim is to develop plans, projects and access to tax incentives linked to telephony, health, housing, among others.

From this perspective, the common interest of urban services managed in a participatory manner with municipal entities and civil society is a *sine qua non* condition for

24

consolidating the region. Focusing on a regional scale, the current configuration shows an intrinsic link between the municipalities of the Big Island and a polarization of neighbouring municipalities that fall within its area of influence.

Table 02: Determinants and characteristics of the RMGSL municipalities.

MUNICIPALITIES	FEATURES
1. San Luis; 2. Sâo José de Ribamar; 3. Paço do Lumiar;	Polarizing cities with a concentration of industrial and service activities with intense commuting.
4. Alcântara;	Strategic geographical location (Alcântara Launch Center - CLA) and significant commuting.
2. Fox; 3. Bacabeira; 4. Rosario; 5. Santa Rita;	Municipalities with a growing level of regional economic activity, interconnected with a large flow of cargo and passengers, with growing commuter traffic and the development of strategic federal and state projects.
6. Axixà; 7. Icau; 8. Hills; 9. Cachoeira Grande; 10. President Juscelino.	Municipalities with a low level of regional economic activity and a weak tax base, resulting in reduced investment capacity and dependence on income transfers. High environmental and tourist potential, with growing commuter traffic and plans for strategic federal and state projects.

It is considered that the incorporation of municipalities at different stages of development and with a predominance of the primary sector and incomplete urbanization, under the area of influence of the capital of Maranhão, in Greater Sao Luis, poses challenges for the optimization of socio-spatial dynamics and restructuring (Chart 02).

This scenario, in general terms, converges on the formulation of the new metropolitan configuration established by LCE No. 174/2015, which aims to meet not only the common demands of the municipalities of Ilha do Maranhao (Sao Luis, Sao José de Ribamar, Paço do Lumiar and Raposa), but to promote integration, development and, as its main motto, the reduction of social inequalities in the municipalities of Alcântara, Axixà, Bacabeira, Cachoeira Grande, Icatu, Morros, Rosàrio, Presidente Juscelino and Santa Rita, based on the union between the state government and the municipalities.

Despite the heterogeneity of local urban arrangements formed in relatively different historical periods and processes, similar or converging trends can be identified in current urbanization processes and the possibility of creating new centralities. In these processes, the traditional centers have been losing population and new contingents have been brought in through a process of intra-metropolitan or regional mobility. In all cases, new housing developments are emerging in the areas furthest away from traditional city centers.

One of the characteristics of the urbanization process in this region is the extensive growth from the central areas of the municipalities, located on the main road axes, configuring various points of conurbation and, at the same time, an unfolding of new urbanized areas, successive subdivisions opened up on the outskirts of the consolidated nuclei that make it up, giving rise to discontinuous urbanization, a strong rural exodus and regionalization of activities, generating intense displacement along the access roads (MASULLO; LOPES, 2016).

In this way, the municipalities in the immediate vicinity of Sao Luis, such as Sao José de Ribamar and Paço do Lumiar, have been receiving large migratory flows, becoming metropolitan sub-centers with increasing influence over the municipalities located on the mainland, in the Munim region.

It can be seen that even with the changes implemented in the light of the Metropolis Statute, the RMGSL needs to develop actions that go beyond institutional issues, given that it still has diffuse and fragmented characteristics in terms of social, economic and even political aspects.

Therefore, the new configuration of the RMGSL currently has the conditions to be consolidated through the shaping and implementation of new territorial strategies, based on an *inclusive urbanism*[9] , with the implementation of new perspectives of regional planning and management in a sectorized way based on local and regional specificities.

---

9 The concept of *inclusion*, in contemporary urbanism, has a dual aspect: guaranteeing citizen participation and diversity of representation in urban planning and, through its results, promoting access for all to services, opportunities and the civic and political life of the city (UN-HABITAT: 2015).

SOCIO-ECONOMIC AND ENVIRONMENTAL CHARACTERIZATION

The socio-economic aspects of metropolitan regions must be analyzed qualitatively, because it is there that the population's quality of life can be seen, with access to essential services in the urban environment. Therefore, the metropolitan issue cannot be thought of unilaterally, but it is also necessary to consider the power of articulation and the current regional issue as a determining element.

Flows and interdependence between cities, where different social and economic realities coexist, are increasing all the time. This scenario brings with it major challenges for the different spheres of government, requiring joint action to optimize regional planning and promote public functions of common interest, whether in metropolises or urban agglomerations.

In order to resolve these issues, it is necessary to broaden our knowledge of the territory and provide a basis for drawing up and implementing public policies on different scales. Thus, the analysis of social, political and economic indicators at local and regional level makes it possible to visualize the obstacles to metropolitan governance and to steer the consolidation of urban policies linked to public functions of common interest.

Social System

Economic empowerment is the main objective of large urban centers, while at the same time highlighting existing social inequalities. This situation is linked to the strengthening of a discourse that offers hope of political, social and economic inclusion in the midst of the urban isolation of the periphery within an exclusionary political and economic model.

Population

The growing population demand in the urban area becomes a problem for the public authorities as soon as there is inefficiency in the urban structure, be it linked to mobility, sanitation, health or even public safety. In this sense, peripheralization can be seen in the city's new areas of occupation, where the equation between what is sustainable and bearable results in a loss of quality of urban life (Figure 06).

Figure 06: Map of the RMGSL's population in 2016.

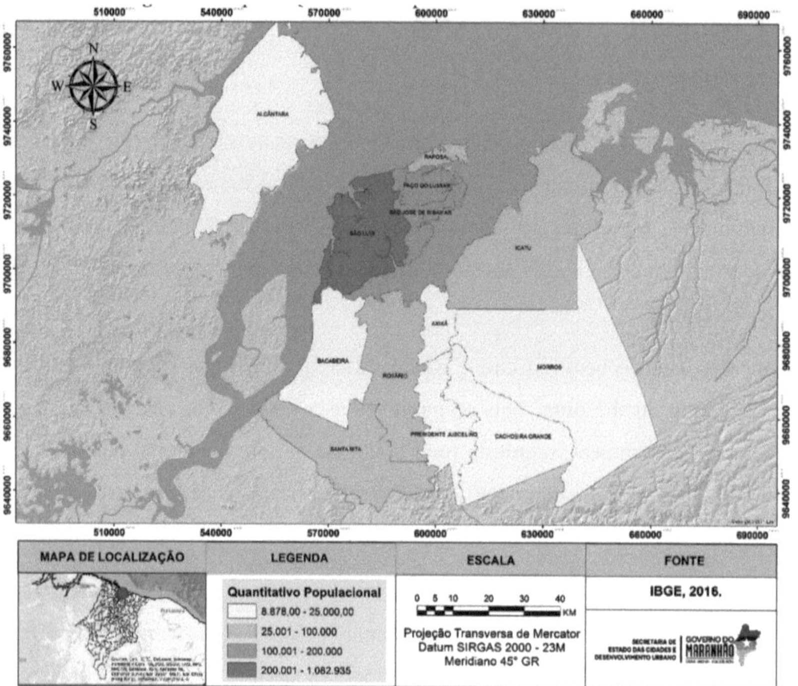

Source: IBGE, 2016.

These impacts are mainly caused by the large-scale growth of urban agglomerations (settlements of more than 5,000 people according to the UN), which, according to estimates, in 1800 only 3% of the world's population lived in cities, 100 years later it rose to 14% and in 1970 38% lived in urban areas, and today more than half of the planet's inhabitants are located in urban agglomerations (SOUZA, 2010).

According to the IBGE (2015), the RMGSL currently has a population of 1,605,305 inhabitants. Approximately 67% of the population is concentrated in Maranhão's capital, while the municipalities of Alcântara, Axixà, Bacabeira, Cachoeira Grande, Icatu, Morros, Raposa, Rosàrio and Presidente Juscelino account for just 14.1% of the region's population (Table 01).

The municipalities of the RMGSL have undergone strong transformations in recent years with the attraction of new companies and investments, which led to considerable population growth between 2000 and 2016. Raposa, Sao José de Ribamar, Paço do Lumiar and Bacabeira were the municipalities with the most significant growth, mainly driven by the installation or forecast of major projects, while in Alcântara there was a reduction in inhabitants between 2010 and 2016, due to the demobilization of the Alcântara Launch Center expansion works.

Table 01: Population evolution of the RMGSL municipalities between 2000 and 2016.

MUNICIPALITY	2000	Growth rate	2010	Growth rate	2016
Alcântara	21.291	3%	21.851	-0,84%	21.667
Axixà	10.142	12%	11.407	4,30%	11.915
Bacabeira	10.516	42%	14.925	11,30%	16.812
Big Waterfall	7.383	14%	8.446	4,90%	8.878
Icatu	21.489	17%	25.145	5,70%	26.651
Hills	14.594	23%	17.883	6,50%	19.116
Paço do Lumiar	76.188	38%	104.881	12,60%	119.915
President Juscelino	10.693	8%	11.541	7,90%	12.532
Fox	17.088	54%	26.327	13,20%	30.304
Rosario	33.665	18%	39.576	5,10%	42.016
Santa Rita	24.922	30%	32.366	11,50%	36.556
Sâo José de Ribamar	107.384	52%	163.045	7,40%	176.008
San Luis	870.028	17%	1.014.837	6,30%	1.082.935
RMGSL	1.225.383	22%	1.492.230	7%	1.605.305

Source: Adapted from IBGE, 2016.

It should be noted that between 2014 and 2015 there was a 1.83% increase in the RMGSL's resident population, of which the region's largest population growth was centered in the municipalities of Sao José de Ribamar and Sao Luis with 11,342 and 9,696 inhabitants respectively. Between 2015 and 2016, there was a 0.94% increase in the population of the RM. This process identifies a situation in which the region suffers from the impacts generated by the implementation of services and projects that affect everyone, without integrated policies to solve these problems.

This is the case, for example, with the Minha Casa Minha Vida (My House My Life) program in Sao José de Ribamar and in the rural areas of Sao Luis, which, on the one hand, is the result of the process of land appreciation that decentralizes popular housing, and on the other, accentuates the lack of services and infrastructure, demonstrating the need to implement mechanisms to integrate the metropolitan dynamic and manage the urbanization process.

Urbanization rate

The advance of urbanization intensifies changes in the environment. Influenced by the dynamics of the city, this system reflects the process of production and reproduction of space based on a logic of distribution and consumption (LIPIETZ, 1979; CORRÊA, 1996). The percentage of urbanization shows and identifies the proportion of the population living in urban areas and those living in rural areas (Figure 07).

According to IBGE (2010), the RMGSL has an urbanization rate of 63.37%, with the centralized conurbed area in the municipalities of Ilha do Maranhao with Sao Luis, Paço do Lumiar and Raposa reaching 94.45%, 74.9% and 63.33% respectively. However, 40% of the municipalities in the RMGSL are predominantly rural and have a low percentage of urbanization, such as Alcântara, Bacabeira, Cachoeira Grande, Icatu, Morros and Presidente Juscelino.

Figure 07: Map of the Urbanization Rate of the RMGSL in 2010.

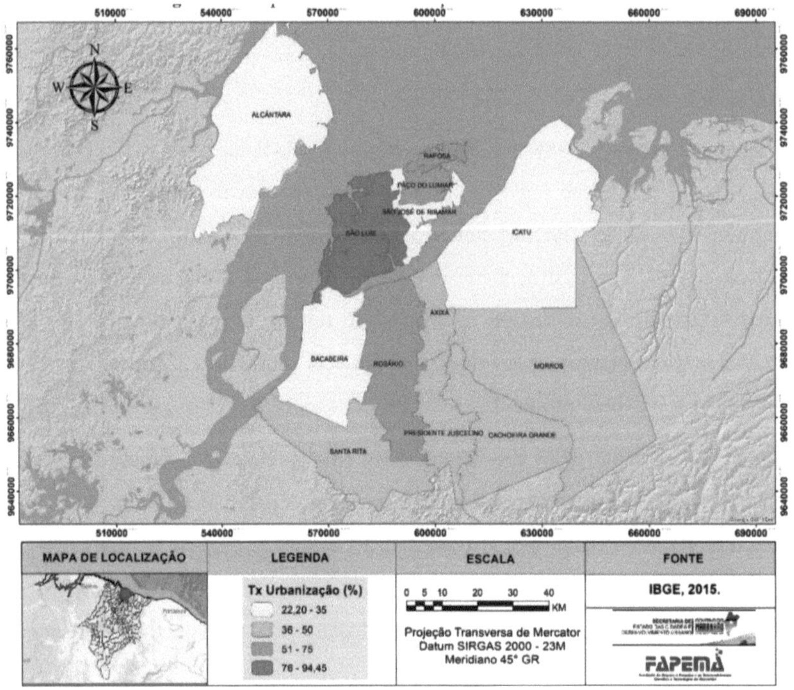

Source: MASULLO; LOPES, 2016.

Road infrastructure

The growth of functionally articulated urban centers influences territorial organization on various scales, reflecting the integration of productive, financial and socio-cultural systems.

This scenario demonstrates the need to reorient the policies that structure the axes of existing Centralities in the region, taking advantage of the degree of road and maritime interconnection, as well as the proximity between municipal seats, which should be seen as an important condition for optimizing management on a regional scale (Figure 08).

Figure 08: Map of the region's air, rail, water and road structures.

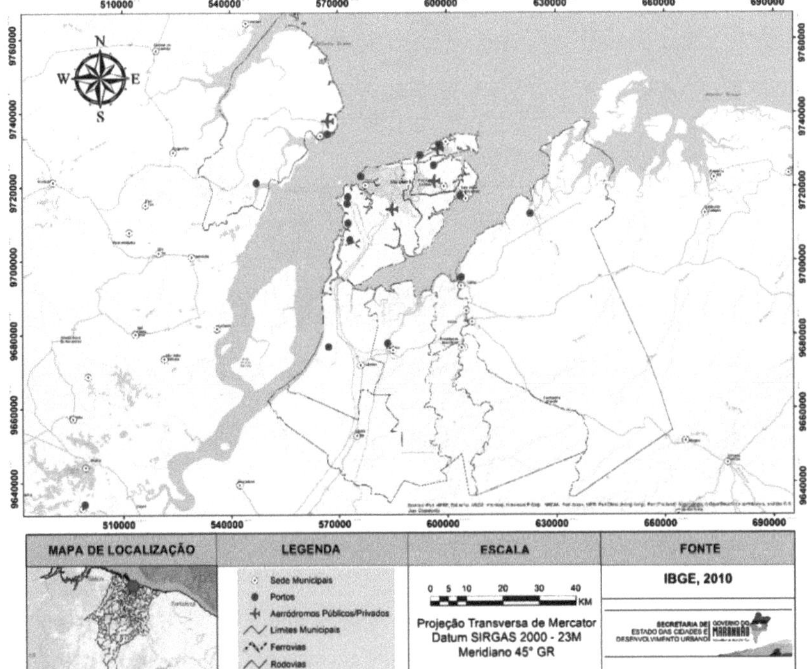

Source: MACROZEE, 2013; IBGE; 2015.

The RMGSL is connected to the interior of the state and to the neighboring states of Parà, Tocantins and Piaui by a railway line, which facilitates agricultural and mineral flows. By road, it is connected by the BR-135, MA 201; MA 202; MA 203; MA 204; MA 402 and MA 020, while by air, it has the Marechal Cunha Machado International Airport, as well as several ports that serve both to transport production and passengers, such as the Port of Itaqui and the Praia Grande Port Terminal (Table 02).

31

Location conditions access and barriers, represented by accessibility and the power of mobility that impacts society at different levels and scales, generating great demands for more specialized mechanisms, due to the low level of network coverage (OLIVEIRA, 2008). Underlying this system of (re)production and consumption of space is the control of travel times and the quality of services, which are fundamental elements and an integral part of the spatial practices of social agents.

Table 02: Distances by access road between the capital of Maranhão and the municipalities of the RMGSL.

Municipalities	Distance between Municipal Boundaries	Distance between Municipal Headquarters	Access routes
San Luis	-	-	MA (201); MA (202); MA (203); MA (204); BR 135.
Alcântara	22 km	22 km	Maritime
Axixà	61 km	91 km	MA 402
Bacabeira	-	53 km	BR 135
Big Waterfall	73 km	76 km	MA 020
Icatu	78 km	109 km	MA 402
Hills	67 km	97 km	MA 402 and MA 020
Paço do Lumiar	3 km	27 km	MA 202 and MA 204
President Juscelino	61 km	117 km	MA 020
Fox	5 km	28 km	MA 203 and MA 204
Rosario	21 km	20 km	MA 402
Santa Rita	48 km	78 km	BR 135
Sâo José de Ribamar	-	28 km	MA 201 and MA 204

Source: Google Earth, 2016.

Housing Deficit

The housing deficit is made up of four variables: inadequacy, replacement, cohabitation and increase. The Deficit of Inadequacy encompasses households in need of adapting existing housing - of a land nature (irregularity or lack of ownership), urban (insufficient or inappropriate infrastructure networks, existence of risk areas, lack of green areas, social facilities) or the building (either because it lacks safety, health and comfort conditions, or because it doesn't have enough internal space for each of the four types of function - sleeping, sanitizing, preparing food and living) (IBGE, 2010).

The replacement deficit is made up of homes that are rustic, built with unsuitable materials or located in risky areas and need to be replaced. As for cohabitation, this is made up of people or families who live in the same home but wish to abandon the cohabitation situation, and finally, the increase deficit occurs when new housing needs to be built to meet demographic growth.

According to the Joao Pinheiro Foundation (2010), the estimated Brazilian housing deficit in 2010 is 6.273 million homes, of which 5.180 million, or 82.6%, are located in urban areas. This figure corresponds to 11.1% of the country's permanent private housing stock, with 10.8% in urban areas and 12.9% in rural areas. The Southeast is where most of these shortages are to be found, with 2.335 million, 37.2% of the total. The Northeast follows suit with 34.2% or 2.144 million households. The MR's make a large contribution to this scenario, accounting for 29.6%, or 1.855 million of the total Brazilian deficit.

Graph 01: RMGSL Housing Deficit in 2010.

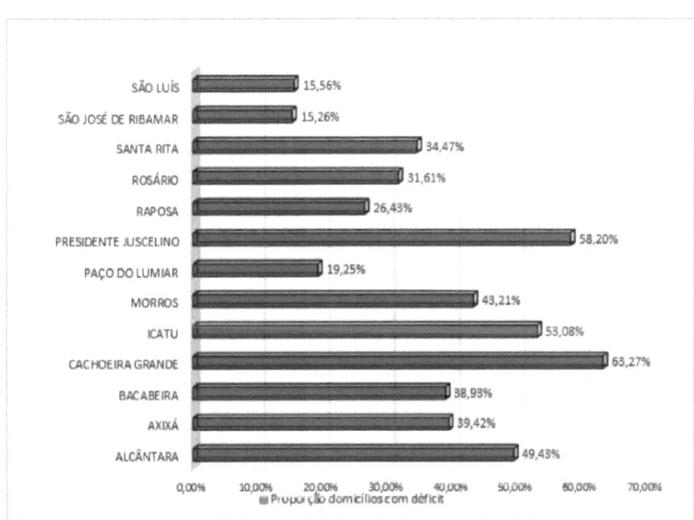

Source: IBGE, 2010.

In Maranhao, the housing deficit stands at 27.3%, while the RMGSL has even more problematic indicators, given that of the 398,785 homes in the region, 33.42% are identified as having a lack of infrastructure, land problems, housing in risk areas and/or cohabitation (Graph 01) (IBGE, 2010).

The highest percentages of deficit in the RMGSL are located in the municipalities of Cachoeira Grande, Presidente Juscelino, Alcântara and Morros, however, it should be noted that quantitatively the largest housing deficits are located in Sao Luis and Sao José de Ribamar. Together, these municipalities account for around 2/3 of the RMGSL's housing deficit.

Human Development Index - HDI

The characterization of the social system involves the analysis of synthetic indicators such as the Municipal Human Development Index (IDHM), calculated every ten years by the Institute for Applied Economic Research (IPEA) and the Joao Pinheiro Foundation (FJP), for states and municipalities, using Census data, with a methodology adapted from the Global HDI.

The MHDI is made up of three dimensions: Income, Longevity and Education. The HDI Income has the per capita income indicator, while the HDI Longevity uses life expectancy at birth, and the HDI Education combines two sub-indices (schooling and school attendance) (IMESC, 2015) (Graph 02).

The MHDI shows a significant evolution of the RMGSL between 2000 and 2010. Based on an average weighted by the population, it is possible to identify that in 2000 the region had an MHDI of 0.617, while in comparison to 2010 there was a considerable increase in the development of the municipalities integrated into the RMGSL, reaching the mark of 0.735, this scenario was driven mainly by the great advance registered in the education indicators.

Graph 02: Comparison of the RMGSL's MHDI between 2000 and 2010.

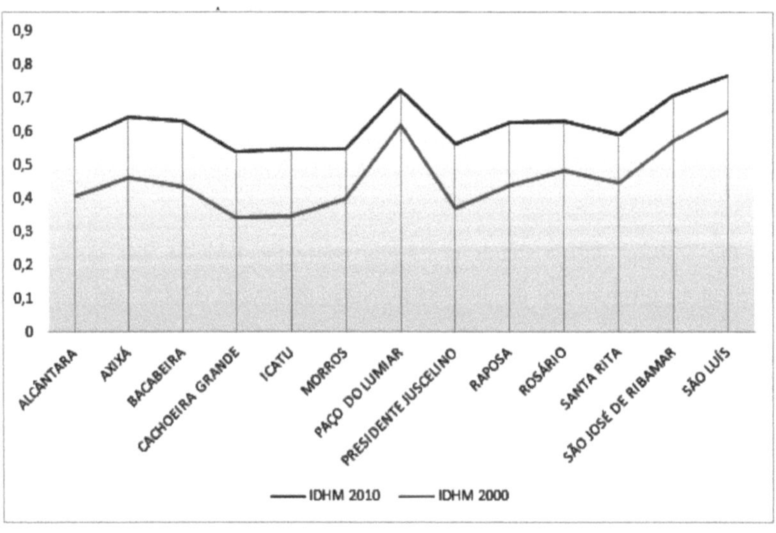

Figure 09: Map of the RMGSL's MHDI in 2010.

It can be seen that the highest Human Development indices are located in the municipalities of Ilha do Maranhao (Figure 09), more precisely in Sao Luis, Paço do Lumiar and Sao José de Ribamar, registering 0.768, 0.724 and 0.708, respectively, while the lowest are in the municipalities of Presidente Juscelino (0.563) and Cachoeira Grande (0.537).

HDI Longevity

In relation to the health dimension of the MHDI, there is the Longevity MHDI, which uses the life expectancy at birth indicator. This indicator is extremely important because it summarizes the level and structure of mortality in a population. The RMGSL has a Longevity HDI of 0.800. It can be seen that all the municipalities have a percentage above 0.700. Among the municipalities with the highest percentages is the Capital with 0.813, followed by Paço do Lumiar with 0.796, while the municipality of Morros has the lowest HDI_L in the RMGSL with 0.706 (Graph 03).

Graph 03: Longevity MHDI of the RMGSL in 2010.

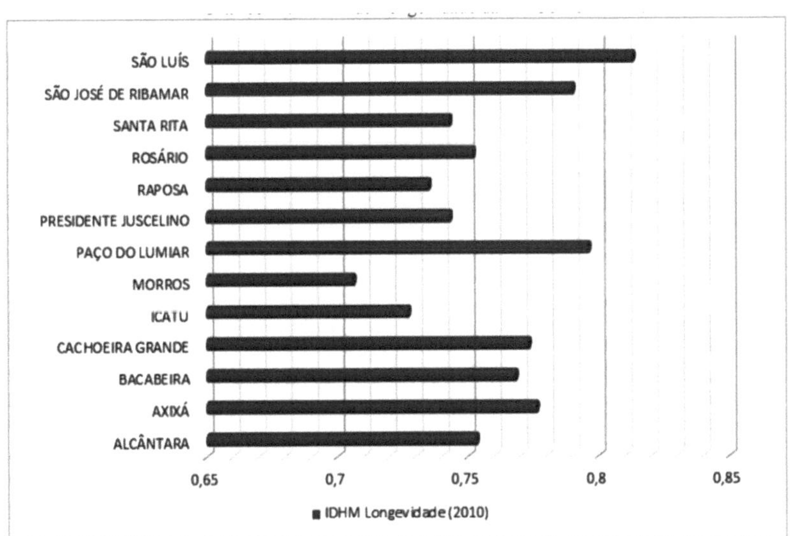

Despite the significant MHDI_L values, there is a need for greater integration and expansion of services linked to both public and private health facilities, given that in the RMGSL, there is a large concentrated volume of infrastructure in the hub city, this factor generates two situations that directly affect a range of people who need this infrastructure.

The first consequence is the agglomeration of people who come to the capital in search of public services or private care, whether it be low or high complexity hospital care: this service is offered precariously by several neighboring municipalities far from Sao Luis, which converges in the dispute for care in the city's hospital network, added to the already high demand from its inhabitants.

The second effect is the precarious basic infrastructure in the municipalities that make up the RMGSL. This scenario causes a decrease in the quality of life of the population, even with the great concentration in the hub city, because in regional terms the fact persists that the municipality of Sao Luis still has a low supply of services, compared to the Metropolitan Regions of the Northeast (Table 03).

Table 03: Number of Health Establishments in the RMGSL.

MUNICIPALITIES	No. of Health Establishments
Alcântara	16
Axixà	15
Bacabeira	II
Big Waterfall	6
Icatu	17
Hills	15
Paço do Lumiar	37
President Juscelino	8
Fox	11
Rosario	20
Santa Rita	27
São José de Ribamar	53
San Luis	1.008
RMGSL	**1.244**

Source: Ministry of Health, 2016.

HDI Education

With regard to education, it is seen as one of the elements that can contribute to an integrated metropolitan project, since the qualification of the population adds investments that can stimulate regional productive restructuring and open up new employment fronts for the population.

The dimension linked to the Education HDI has two sub-indices: schooling (the percentage of people aged 18 or over with completed primary education) and school attendance (the percentage of children aged 5 to 6 in school, the percentage of adolescents aged 11 to 13 in the final years of primary education or with completed primary education and the percentage of adolescents aged 15 to 17 with completed primary education) (IMESC, 2015).

Graph 04: Education MHDI of the RMGSL in 2010.

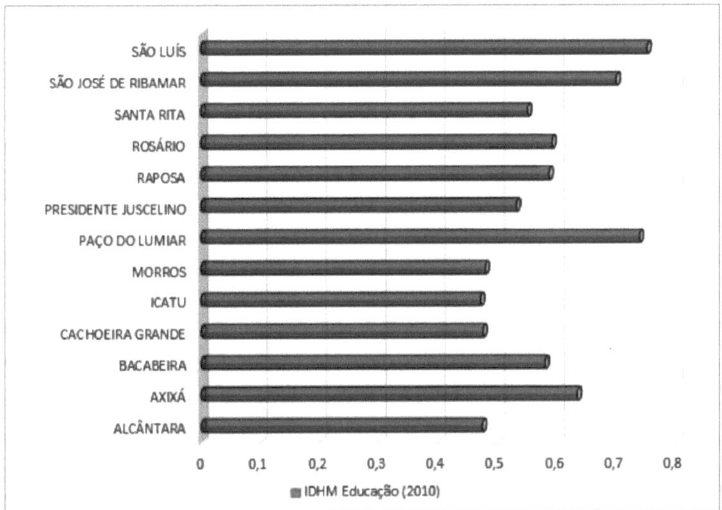

Source: IPEA, 2013.

When analyzing education specifically, one can see the great evolution of the indicator, which in 2000 reached 0.531 and in 2010 reached 0.716, the MHDI_E was the indicator that most influenced the advance of the MHDI of the RMGSL with the substantial improvement of its sub-indices. Once again, the highest indices were identified in the municipalities of Sao Luis and Paço do Lumiar with 0.752 and 0.739 respectively, while Icatu with 0.472 and Alcântara with 0.475 have the lowest percentages in the RMGSL.

Figure 10: Illiteracy rate in the RMGSL in 2010.

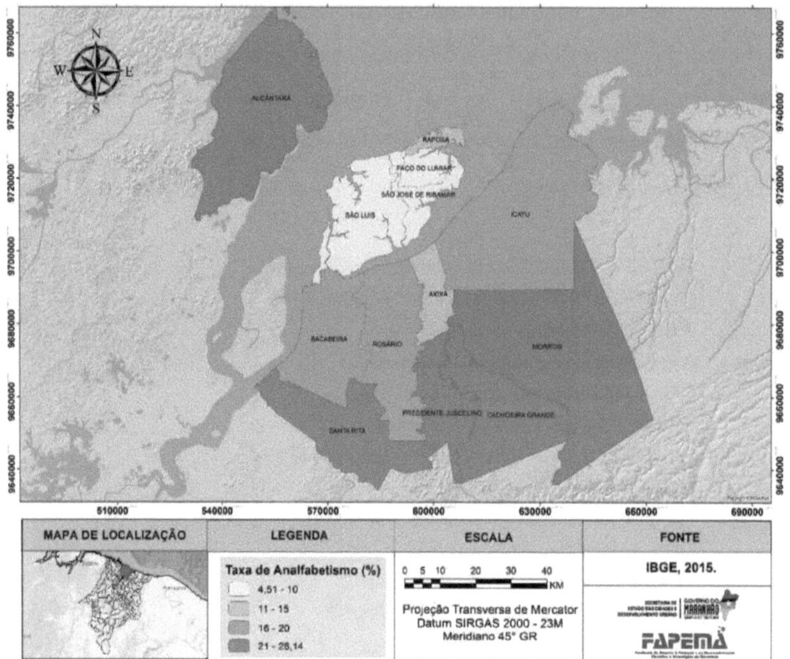

Source: IBGE, 2010.

The illiteracy rate was also used as an indicator to portray the living conditions of the population living in the RMGSL, given that the illiteracy rate in Maranhao reaches 20.9% according to IBGE (2010). In the RMGSL, there are municipalities with illiteracy rates higher than the state average, such as Alcântara, Morros, Cachoeira Grande, Santa Rita and Presidente Juscelino, with percentages ranging from 20% to 28%. Although still unsatisfactory, the lowest percentages are found in Sao Luis, Paço do Lumiar and Sao José de Ribamar, with less than 6% of the population considered illiterate (Figure 10).

The educational levels of the RMGSL in most of the municipalities that make it up are still low compared to the levels of the country. This situation demonstrates the precariousness of the education offered to the population, as well as the lack of basic school infrastructure in most of the schools managed by the municipalities (Table 04).

The process of metropolitanization of a region that encompasses various economic, social and environmental factors is uniquely complex, because between political metropolitanization and the functional implementation of the metropolis, there is a considerable distance that hinders the desire to solve problems of common interest and the urban policies necessary for the intergovernmental management of an RM.

Table 04: Education levels for basic indices in 2010.

Spatialities		% of employed with complete primary education - 18 years and over (2010)	% of employed with complete secondary education - 18 years and over (2010)	% of employed with a university degree - 18 years or older (2010)	% of the employed with full 11 to 14 years (2010)	Illiteracy rate - 15 years and over (2010)	Illiteracy rate - 15 and with complete primary education (2010)	% aged 18 and over (2010)	Expected over years of study (2010)
	Brazil	62,29	44,91	13,19	3,24	9,61	54,92	9,54	
Alcantara		48,99	29,57	4,54	6,26	22,44	37,94	8,21	
	Axixa	49,7	32,85	6,33	3,54	16,37	49,37	9,61	
Bacabeiia		48,12	26,45	3,28	5,08	18,3	44,35	9,44	
	Big Waterfall	29,94	19,36	3,21	13,29	30,41	30,58	8,74	
-U	Icatu	38,38	24,96	5,18	3,89	22,13	30,71	9,75	
	Monos	34,33	23,12	4,87	17,06	29,78	35,64	8,17	
Paca do Lumiar		76,62	59,72	7,27	2,96	5,76	73,02	9,84	
	President Juscelino	43,82	30,11	4,19	9,22	28,34	38,71	8,53	
Rapasa		49,46	30,58	3,35	5,31	15,27	45,67	9,67	
	River rose	50,79	35,78	4,19	6,46	18,1	47,29	9,84	
Sauta Rita		46,43	29,01	3,4	7,38	22,12	39,53	9,54	
	Sào José de Ribamar	70,78	52,84	7,24	2,43	6,7	66,93	10,21	
Sao Luis		78,07	61,64	15,54	2,37	4,65	73,45	9,84	

Foute: PNLID, 2010

Economic System

Gross Domestic Product - GDP

The economic aspects show the strengths and weaknesses of the current model, which is reproduced within a dichotomous and spatially segregated space. The Gross Domestic Product (GDP) of an MR qualifies the flows of capital that circulate and are produced in the cities, demonstrating economic strength or subsistence.

In the state of Maranhao, there are municipalities with economic growth rates above the Northeast region or even above the levels expected for the country, as seen in the analysis of GDP per municipality. The RMGSL's GDP rose steadily between 2010 and 2013, with growth of around 11% and 14% between 2011 and 2012 respectively (Graph 05).

Graph 05: Gross Domestic Product of the RMGSL between 2002 and 2014.

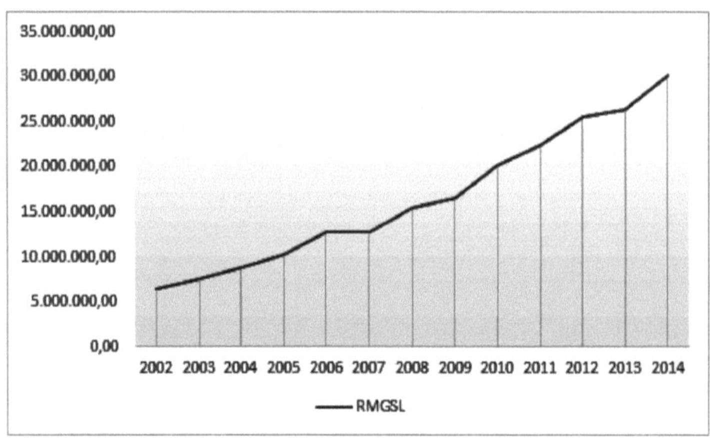

Source: IMESC, 2016.

According to IMESC (2016), the RMGSL represents 39.4% of Maranhao's GDP, accounting for R$ 30.2 billion reais, 87% of which was concentrated in the Maranhão capital in 2014. The rest of the RMGSL's GDP is divided between the municipalities of Sao José de Ribamar, Paço do Lumiar and Bacabeira, which account for 9% of the GDP, while 4% is redistributed between Alcântara, Axixà, Cachoeira Grande, Icatu, Morros, Raposa, Rosàrio, Presidente Juscelino and Santa Rita (Figure 11).

Figure 11: Map of the RMGSL's GDP in 2014

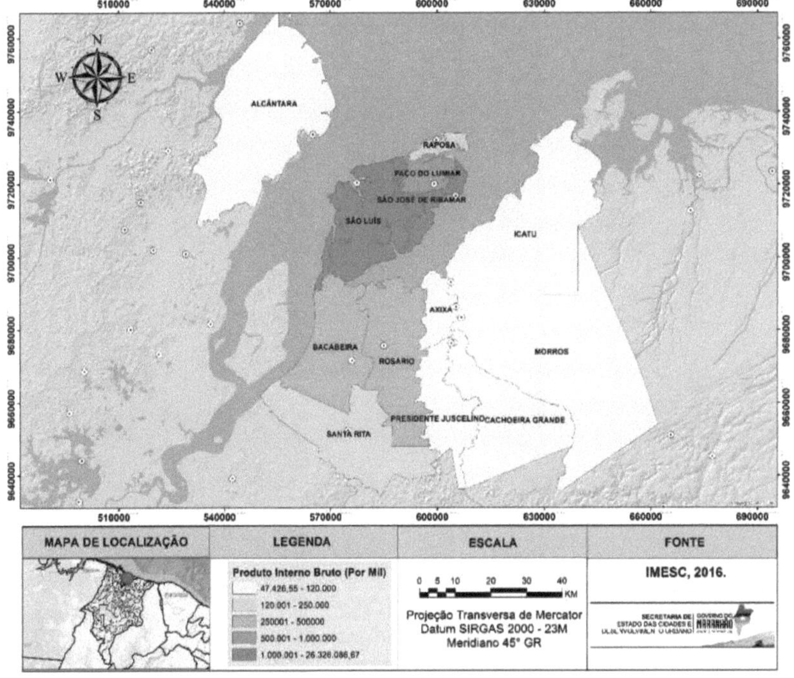

Source: IMESC, 2016.

It is clear to see that the wealth generated in the RMGSL is concentrated in Sao Luis, which attracts a large influx of migrants from neighboring municipalities that act mainly as dormitory towns and second homes, or even from more distant municipalities in search of employment opportunities. This scenario of economic centralization in the capital increases social inequalities and reduces the quality of life in the region.

Making up the GDP is the Value Added in the Services sector, where its percentage is highly concentrated in the municipalities of Sao Luis, Sao José de Ribamar and Paço do Lumiar, especially in the areas of Public Administration and Commerce, while municipalities such as Axixà, Presidente Juscelino and Cachoeira Grande have the lowest percentages.

This comparison shows the great disparity that exists in the RMGSL, where the capital concentrates 88% of the region's Value Added in Services (Figure 12).

Figure 12: Map of RMGSL's VA of Services in 2014

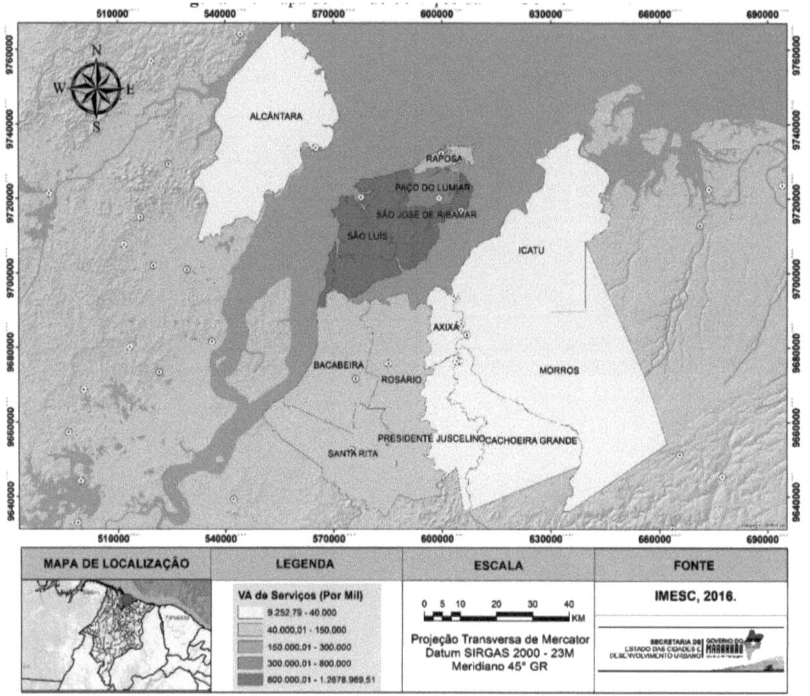

Source: IMESC, 2016.

In terms of agricultural value added, the highest percentages are found in the municipality of Presidente Juscelino, followed by Sao Luis, Icatu and Rosàrio. These municipalities have a growing demand for agriculture, with a higher value added than Sao Luis and Sao José de Ribamar, while Morros and Paço do Lumiar have the lowest percentages (Figure 13).

The industrial sector is centered on extraction and transformation. Of particular note are VALE and ALUMAR, which, underpinned by the port complex formed by the terminals at Itaqui, Ponta da Madeira and ALUMAR, have aggregated their sectoral activities in other regions of the state, highlighting the yogic aspects of the outflow of production in the state and the metropolitan region.

Figure 13: Map of the Agricultural VA of the RMGSL in 2014.

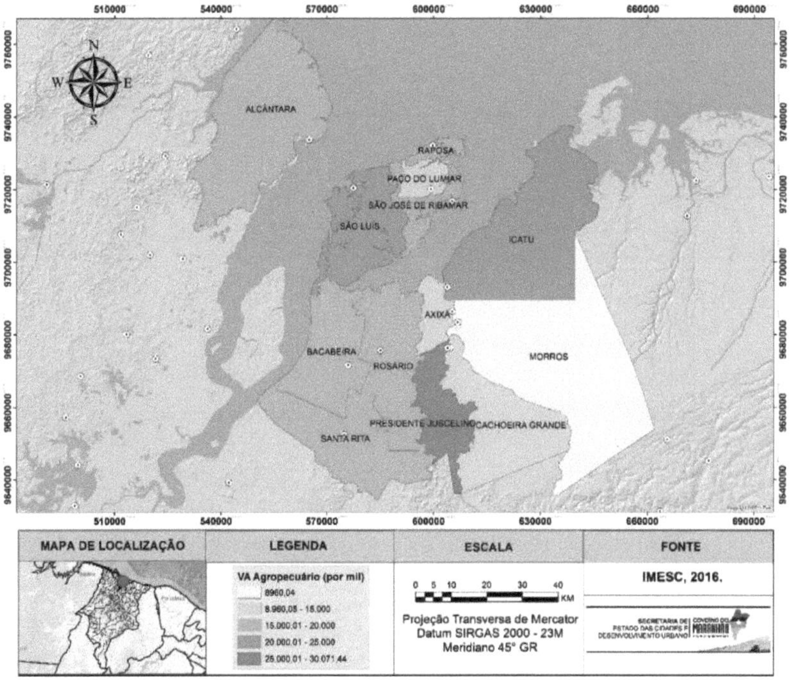

Source: IMESC, 2016.

Civil construction is also very important, especially in Maranhão's capital, which accounts for 90% of the RMGSL's Industrial Value Added, followed by Sao José de Ribamar, Paço do Lumiar and Bacabeira. The municipality of Rosàrio also stands out for its mineral extraction potential, mainly extracting substances such as granite, clay and sand.

In the RMGSL, there are municipalities with rural characteristics and low industrial activity, such as the municipalities of Axixà, Presidente Juscelino and Cachoeira Grande (Figure 14).

Figure 14: Map of the Industrial VA of the RMGSL in 2014.

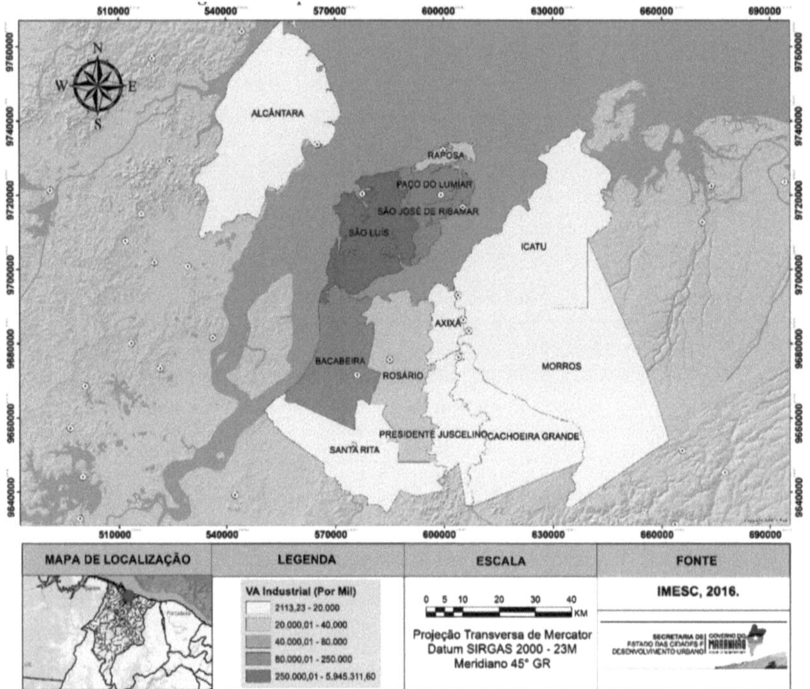

Source: IMESC, 2016.

This accumulative character is the result of the organization of the territory, directed by the existing centralities and hierarchies, notably marked by historical issues. In metropolitan regions, the Centralities base the forms of reproduction of urban life on the use value and appropriation of qualities and contents that are different from the central locations, hierarchized in a network (SERPA, 2012).

In this way, the municipality of Sao Luis concentrates the main urban services in the RMGSL, meeting existing demand, and therefore ends up exerting a centripetal force in the urban hierarchy, linking its urban fabric to the other municipalities in the region. However, areas of common interest to the municipalities of the RMGSL should be the target of shared management in order to create new centralities aimed at reducing the existing concentration of production and local and regional inequality.

MHDI Income

The per capita income of Brazilians has increased significantly in recent years. The RMGSL follows this growth, as shown by the region's Income HDI, which in 2000 registered 0.618, while in 2010 there was a significant increase in per capita income, reaching 0.693 (Graph 06).

Graph 06: RMGSL's Income MHDI in 2010.

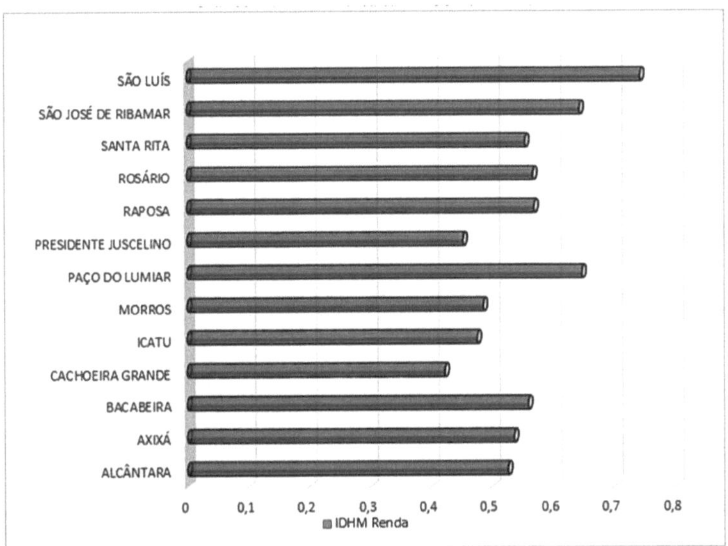

Source: IPEA, 2013.

The municipalities of Rosàrio and Bacabeira have seen the greatest growth in the RM's Income HDI. However, there is still great inequality between the municipalities of the RMGSL, given the great concentration of income in Sao Luis, Paço do Lumiar and Sao José de Ribamar, while the municipalities of Presidente Juscelino and Cachoeira Grande still have low income indicators with 0.452 and 0.422 respectively.

Population in the Labor Force

According to the IBGE (2010), Maranhao's Labor Force Population (TFP) is 2,585,063 people, with 27% located in the municipalities that make up the RMGSL. However, when formal jobs via the CLT and statutory system are evaluated, it can be seen that only 28.58% of the state's TFP is covered, in addition to the fact that 52.68% of Maranhão's formal jobs are concentrated in the RMGSL (Table 05).

Table 05: Population in the Labor Force and CLT and Statutory Contracts.

MUNICIPALITIES	LABOR STATUTORY	AND TFP	(%)

	CONTRACTS		
Alcântara	1.626	7.692	21,13%
Axixà	465	4.255	10,92%
Bacabeira	2.690	5.812	46,28%
Big Waterfall	162	2.382	6,80%
Icatu	600	7.352	8,16%
Hills	1.096	6.759	16,21%
Paço do Lumiar	5.788	47.547	12,17%
President Juscelino	468	3.052	15,33%
Fox	1.957	10.382	18,84%
Rosario	2.003	16.029	12,49%
Santa Rita	1.384	12.384	11,17%
Sao José de Ribamar	16.043	71.342	22,48%
Sao Luis	354.124	502.567	70,46%
RMGSL	388.406	697.555	55,68%

Source: IBGE, 2010; RAIS, 2014.

The economically active population in the RMGSL totals 697,000 people, 55.68% of whom work under the CLT or statutory system. 91% of these are located in São Luis, followed by Sâo José de Ribamar and Paço do Lumiar in quantitative terms, But it should be noted that the municipality of Bacabeira has a high percentage of formalized jobs concentrated in the industrial sector, with 46% of its population working under the CLT or statutory system, while Cachoeira Grande and Icatu have percentages of less than 10%.

These indicators help us to understand the essential elements of the local and regional situation based on socio-spatial differentiation. In order to counter this reality, the production of space and the development of the process of differentiation must be understood, with a view to developing and implementing territorial strategies capable of articulating, integrating and reducing existing social and economic inequalities.

Environmental System

Areas of Environmental Relevance

Metropolitan regions are fragmented habitats, which represents one of the greatest threats to local and regional biodiversity (TABARELLI; GASCON, (2005). The main consequences of this reality are the isolation of remaining formations and populations, changes in gene flows, intensification of intra- and inter-specific competition, changes in the structure and quality of habitats, species extinctions and loss of biodiversity (METZGER, 1998,

PRIMACK; RODRIGUES, 2001).

It is therefore extremely important to define parameters in order to prescribe management actions aimed at conserving, maintaining or expanding the biodiversity of forest fragments (NUNES et al, 2005).

Land use planning, taking into account the spatial distribution of forest remnants, has become an important tool for proposals aimed at minimizing the impacts caused by natural habitat fragmentation (MUCHAILH, 2007).

The RMGSL's large urban agglomeration is home to the Conservation Units - UC's of the Maracana Environmental Protection Area; the Rangedor Ecological Station; the Itapiracó Environmental Protection Area; the Upaon-Açu - Miritiba - Alto do Rio Preguiças Environmental Protection Area; the Baixada Maranhense Environmental Protection Area; the Reentrâncias Maranhenses Environmental Protection Area and the Bacanga State Park (Chart 03; Figure 15).

Chart 03: RMGSL Conservation Units.

Conservation Unit	Decree	Degree of Protection
Upaon-Açu - Miritiba - Alto Preguiças State Environmental Protection Area.	No. 12.428 of June 5, 1992	Sustainable Use
Reentrâncias Maranhenses Environmental Protection Area	No. 11.900 of June 11, 1991 and reissued on October 5, 1991	Sustainable Use
Bacanga State Park	No. 7.545 of March 2, 1980	Comprehensive Protection
Rangedor State Park	No. 21.797 of December 15, 2005, amended by Decree No. 23.303 of August 7, 2007 and Bill No. 321/2015	Comprehensive Protection
Maracanã Environmental Protection Area	No. 12.103 of October 1, 1991	Sustainable Use
Itapiracó Environmental Protection Area	No. 15.618 of June 23, 1997	Sustainable Use
Baixada Maranhense Environmental Protection Area	No. 11.900 of June 11, 1991 and reissued on October 5, 1991	Sustainable Use

Source: Adapted from MASULLO, 2015.

Approximately 63% of the RMGSL is covered by Conservation Units, concentrating 25% of the RMGSL's inhabitants. The vast majority of these are units with sustainable use characteristics, which allow a certain degree of human occupation and exploitation of natural resources, with the aim of using ecosystems in a direct and sustainable way.

With the expansion of these areas, it is necessary to set up comprehensive and integrated planning, from a systemic viewpoint, which will make it possible to monitor and implement actions to preserve PAs in Metropolitan Regions.

Thus, there is a need to expand knowledge of biodiversity on a regional scale, based on the principle that even with the large expansion of areas theoretically protected by UCs, the RMGSL does not have the necessary conditions to guarantee the conservation of its ecosystems, which continue to be threatened by the increase in deforestation and fires, leading to the fragmentation of habitats, the destruction of landscapes and the consequent extinction of animal and plant species (MASULLO; LOPES, 2016).

Figure 15: Map of Use and Occupation of RMGSL Conservation Units

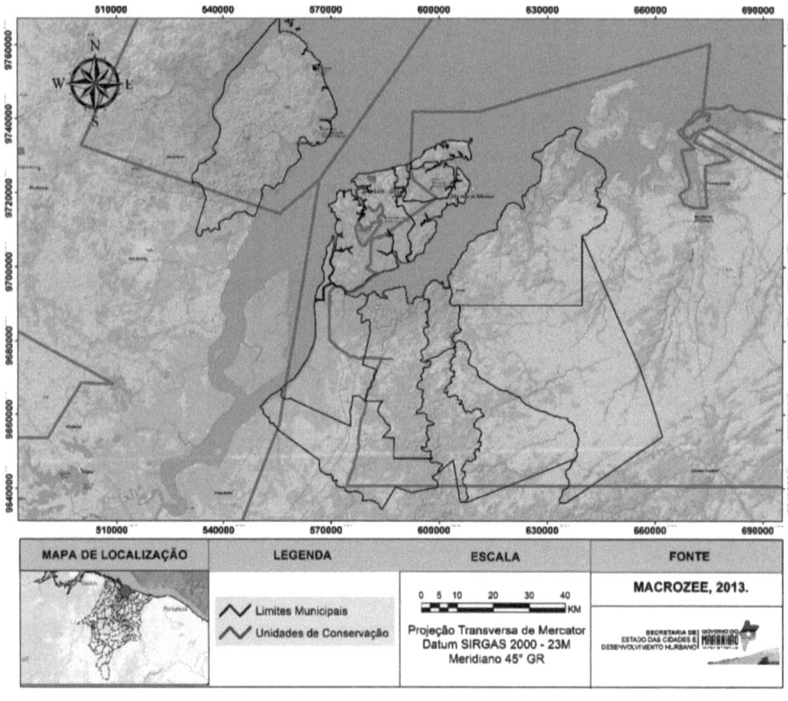

Source: MACROZEE, 2013.

It can be seen that the RMGSL has developed under the influence of Maranhão's capital, with artificialized spaces based on earthmoving techniques and successive landfills of spring areas, mangroves and small watercourses, as well as the emergence of new areas of occupation and new housing developments, causing various impacts on its protected areas. The data from the processing of satellite images corroborates the above statements, showing urban growth and the percentage of preserved areas in the Metropolitan Region's PAs

(Table 06).

The UCs with the highest percentage of vegetation include the Rangedor State Park, Bacanga State Park and Itapiracó APA with 96.8%, 93.5% and 76.1% respectively, while the Baixada Maranhense and Reentrâncias Maranhenses APAs reached percentages of more than 70% of urban area and exposed soil.

Table 06: Comparison of Occupation in the RMGSL PAs

Conservation Units	Vegetation (%)	Urban Area and Exposed Soil (%)
APA Upaon-Açu - Miritiba - Upper Preguiças River	73,1	26,9
Bacanga State Park	93,5	6,5
Rangedor State Park	96,8	3,2
Itapiracó APA	76,1	23,9
Maracana APA	33,2	66,8
Baixada Maranhense APA	11,8	88,2
APA Reentrâncias Maranhenses	29,7	71,3

Source: MASULLO, 2016.

These changes in the ecosystems of the Conservation Units are amplified and come to be underpinned by social and economic factors, which influence the dynamics of the landscape and local and regional susceptibility. It is therefore important in the context of science to study the interdependence between natural and social processes on different temporal and spatial scales, understanding this relationship in an integrated way, supporting environmental planning and management (BECKER, 2010).

Basic Sanitation

Among the public functions of common interest, basic sanitation must be worked on across the board, in accordance with Federal Law No. 11.445/2007, which provides for the universalization of water supply, sewage and rainwater drainage services, as well as garbage collection to guarantee the health of the population. However, in the RMGSL there are serious problems with sanitation, causing harm to society, such as the closure of beaches and an increase in the proliferation of numerous diseases (MASULLO, 2013).

Maranhao has a precarious sanitation infrastructure. According to IBGE (2010), the percentage of households receiving solid waste collection services in Brazil is 87%, this figure is reduced in the Northeast region to 62% and in Maranhao it does not exceed 55%. Waste collection in the RMGSL follows this trend, although it is higher than in the Northeast and the state, with 74.2% of households having regular collection (Figure 16).

Figure 16: Map of the distribution of Solid Waste Collection in the RMGSL in 2010.

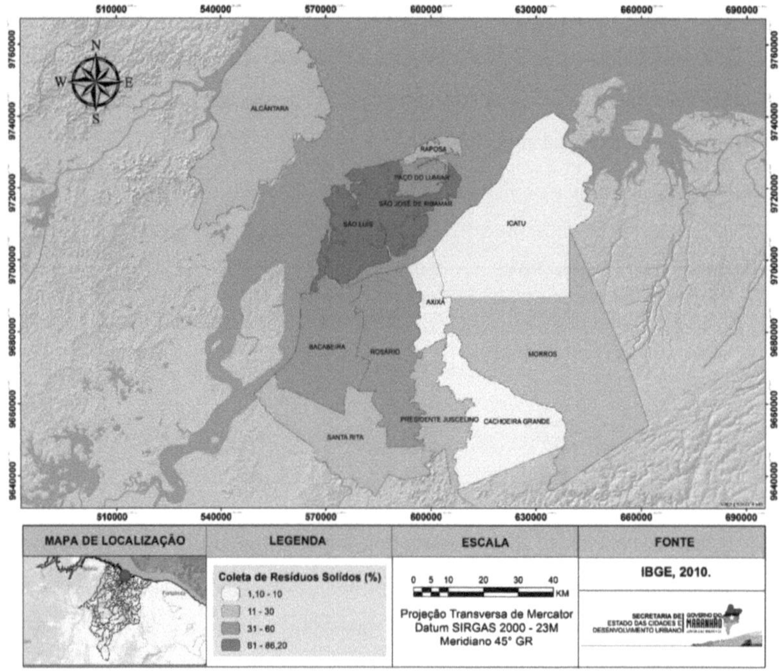

Source: IBGE, 2010.

The highest percentages of waste collected are on the island of Maranhao, with collection services of over 50%. In the case of Sao Luis, 86.2% of households have waste collected. In municipalities such as Axixà, Cachoeira Grande and Icatu, the collection percentages are extremely low, at 1.1%, 2.7% and 8.3% respectively. It is noteworthy that even the municipalities with the highest percentages of households served have the service concentrated in the municipal headquarters, leaving the inhabitants of rural areas unattended. This scenario shows the great need to set up an integrated governance structure that not only increases the number of households with regular collection, but also prioritizes the proper disposal of solid waste, through consortium operation with medium and long-term planning.

In relation to the water supply network, the country has 82% of households served, in the Northeast this figure drops to 68% and in Maranhao it reaches 65.8% (IBGE, 2010). In the RMGSL, 78.8% of households are connected to the general water supply network. Among the municipalities that make up the RM, only Presidente Juscelino has a percentage of less than 50% of assisted households, accounting for only 44% of households with a water supply, while the highest percentages are in the municipalities of Raposa with 85.3% and Sao Luis with 83.2% (Figure 17).

Figure 17: RMGSL Water Supply Distribution.

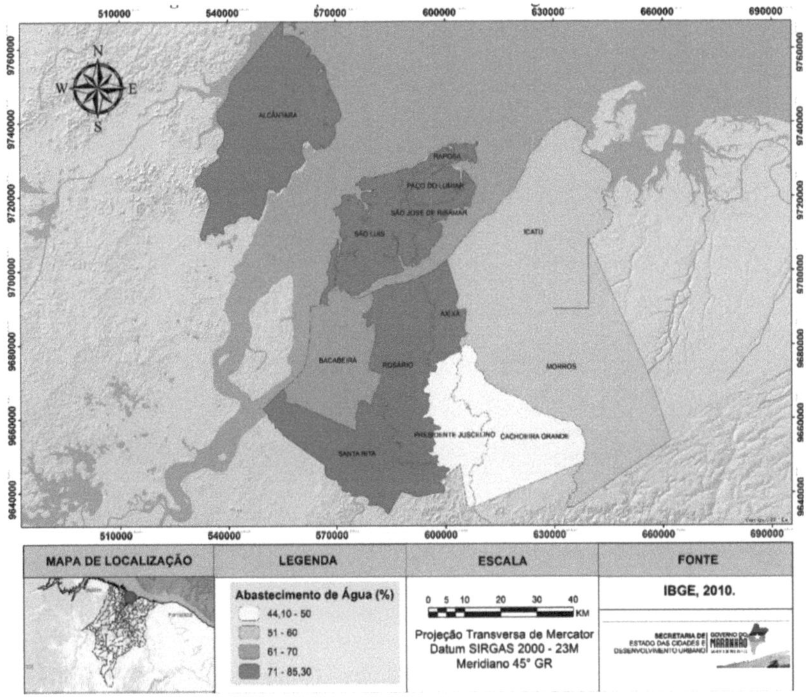

Source: IBGE, 2010.

In Brazil, according to IBGE (2010), 55% of households have a sewage system, while the percentage in the Northeast and Maranhao is much lower than the national figure. The Northeast region has 33.90%, while the state has only 11.60% of households with a sewage system and around 15% with septic tanks. The RMGSL has a much higher percentage than the state and the northeast, with 55.9% of households with sewage services, including sewage systems and septic tanks (Figure 18).

However, it must be stressed that this percentage should be analyzed not only on a regional scale, but also on a municipal level. The highest percentages of households with sewage collection systems and septic tanks are found in the municipalities of Sao Luis, Paço do

51

Lumiar and Sao José de Ribamar, with rates of over 45%, while the lowest percentages are found in the municipalities of Alcântara, Cachoeira Grande and Icatu.

Figure 18: Distribution of sewage disposal in the RMGSL in 2010.

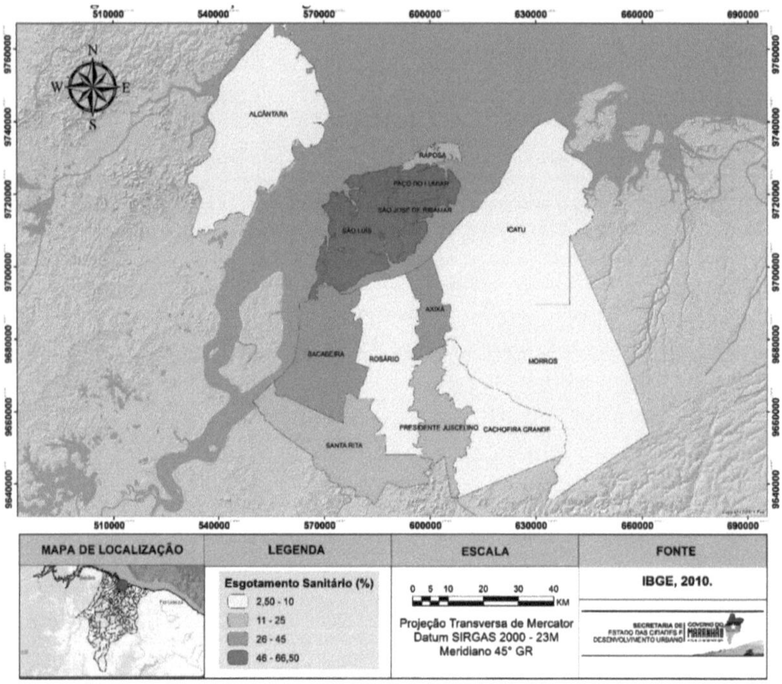

Source: IBGE, 2010.

It should be noted that even in the island's municipalities, sewage disposal is almost entirely concentrated in the urban area, while a significant portion of the population lives with a complete lack of service, both in terms of collection and effluent treatment. This model ends up being replicated in all the municipalities that make up the RMGSL, leading to intensified environmental degradation and a consequent reduction in local and regional tourism.

In order to solve this problem, it is necessary to have a great deal of coordination between the state and municipal authorities, with the aim of drawing up and implementing the RGMSL's Basic Sanitation Plan, with a view to joint and structuring actions that will improve the population's quality of life.

Water Resources

One of the most important issues in the field of urban planning is the need for an institutional framework that adequately represents the urban, territorial and socio-economic planning that characterizes metropolitan regions. From this urban perspective, various issues arise on a supra-municipal scale, requiring a more comprehensive approach. In this sense, the integrated management of water resources is of great importance.

Notably, the municipal administrative sphere is the closest to the social reality, but its scale of action does not allow for a systemic view of the territory in which it is inserted. In the RMGSL, the absence of a clear definition of the duties of local governments stands out, a fact that is aggravated by the reduced budgetary autonomy of the municipalities, making them dependent on financial transfers from the state and federal levels. This makes it difficult or even impossible for them to participate more effectively in the management of various sectors, such as water resources.

The RMGSL has the Mearim, Itapeucuru, Munim and Perià Hydrographic Basins, as well as the Western Coast Hydrographic System and the Maranhão Islands Hydrographic System, which includes the basins on Maranhao Island (Estiva, Ihaùma, Rio dos Cachorros, Tibiri, Itaqui, Bacanga, Anil, Guarapiranga, Jeniparana, Paciência and Praias) (Figure 19).

The Mearim River Basin, with an area of 99.058.6 km^2 (29.84% of the territory), is the largest of all the hydrographic basins in Maranhao. The Mearim River is 930 km long, rising in the Serra da Menina and flowing into the bay of Sao Marcos. Its main tributaries are the Pindaré and Grajaù rivers, covering, among others, the municipalities of Bacabeira and Santa Rita, which make up the RMGSL (UEMA, 2011).

With regard to the economic importance of the Mearim River Basin for the RMGSL, its industrial potential stands out, given its strategic location for transporting production.

Figure 19: River basins that make up the RMGSL

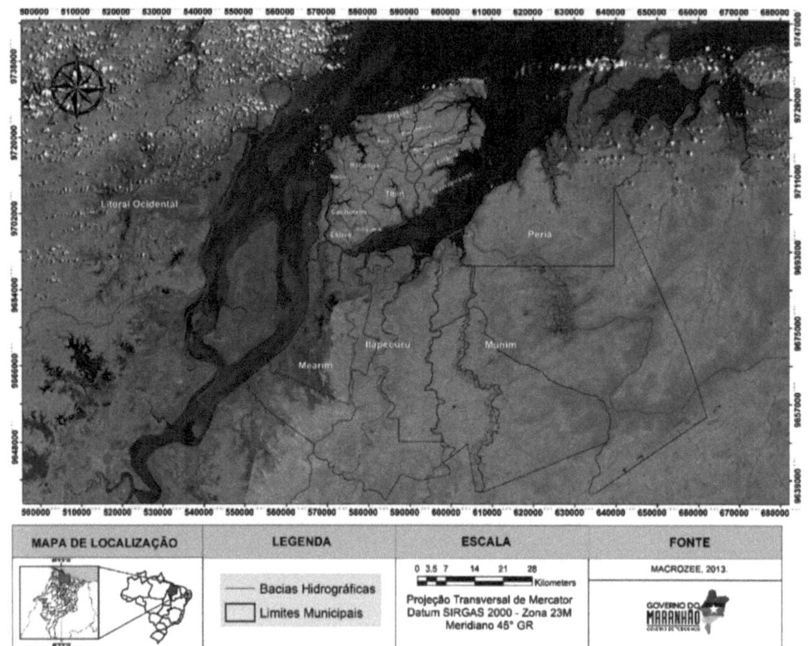

Source: MACROZEE, 2013.

The 1,450 km long Itapecuru River Basin covers an area equivalent to 16.03% of Maranhao, totaling 53,216.84 km^2 , comprising 57 municipalities, with a population of approximately 1,600,000 inhabitants (UEMA, 2011). In the RMGSL, the Itapecuru River runs through the municipalities of Bacabeira, Rosàrio and Santa Rita, flowing into the São José basin. It is currently considered to be of vital importance to the region as it supplies more than 50% of its population.

As for the Munim River Basin, which runs through the RMGSL municipalities of Axixà, Cachoeira Grande, Icatu, Morros and Presidente Juscelino, it covers a population of approximately 320,000 inhabitants over an area of 275 km and 15,918.04 km2, which represents 4.79% of the state's area. In addition to the main river, its many tributaries, such as the Una, Iguarà, Paulica and Preto rivers and the Pirangi, Mocambo, Raiz, da Cruz, Sao Gonçalo and da Mata streams, are also important for tourism and fishing (UEMA, 2011). The Perià, Mapari and Anajatuba rivers form the Perià River Basin, with a population of 64,049 inhabitants, 80 km in length and an area of 5,395.37 km2, equivalent to 1.62% of the territory of Maranhão, comprising the RMGSL municipalities of Icatu and Morros (UEMA,

2011).

The Western Coast Hydrographic System covers an area of 10,226.22 km2, representing 3.08% of the entire territory of Maranhão, with an approximate population of 343,130 inhabitants. It is made up of the drainage areas of the Pericuma, Aura and Uru rivers. Its economy consists of plant extraction and fishing, and it has great potential for the installation of industrial enterprises due to its port facilities.

The Maranhão Islands Hydrographic System is made up of 219 islands, covering an area of 3,604.62 km2, equivalent to 1.09% of the state. It covers 22 municipalities including the capital Sao Luis, Paço do Lumiar, Raposa and Sao José de Ribamar. The system has a population of around 1,349,541 inhabitants, 20.5% of our state's population, and has the highest demographic density in Maranhao (UEMA, 2011). It is also the area with the greatest diversity of economic activities and the highest rate of development.

Therefore, in the RMGSL there are municipalities that are inserted in watersheds of varying length and with different levels of complexity. Thus, there are basins located entirely on the island of Maranhao, with a high population density and high environmental fragility, and extensive basins with low population density, the latter being basically made up of the middle and lower reaches of the river, which refer to the synthesis of the environmental processes developed throughout this drainage network, highlighting erosive and depositional processes (Table 07).

Table 07: Population of the RMGSL by Hydrographic Basin.

WATER BASINS	POPULATION
ANIL	273.664
BACANGA	243.012
RIVER OF DOGS	18.967
ESTIVA	8.077
GUARAPIRANGA	671
INHAÙMA	3.199
ITAPECURU	64.789
ITAQUI	35.780
JENIPARAMA	55.257
WEST COAST	21.652
MEARIM	22.731
MUNIM	50.681

PATIENCE	372.041
PERIÂ	17.455
BEACHES	134.815
SANTO ANTÔNIO	193.332
TIBIRI	45.063

Source: IBGE Microdata, 2010.

In addition to the physical aspects, the number of inhabitants per BH also stands out. This data allows for the definition of strategic actions aimed at regional development. It can be seen that the Paciência, Anil, Bacanga, Santo Antônio and Praias basins have absolute numbers of more than 100,000 inhabitants. These figures highlight the need for actions to preserve areas that are relevant to the preservation of the quantitative and qualitative aspects of these basins. On the other hand, basins with a small number of inhabitants need to be addressed in order to guarantee development through appropriate policies.

At the same time, the municipality is responsible for drawing up, approving and supervising instruments related to land use planning. Tucci (2004, p. 99) presents some factors that make it difficult to apply the concepts of integrated water management in urban areas. These are

Generalized lack of knowledge on the subject: the population and professionals from different areas do not have adequate information on the problems and their causes. Inadequate conception by engineering professionals of system planning and control: a significant proportion of technicians working in urban areas are out of date in terms of their environmental vision and generally look for structural solutions that alter the environment, with an excess of impermeable areas and, consequently, an increase in temperature, flooding, pollution, among others; Sectorized vision of urban planning: the planning and development of urban areas are carried out without incorporating aspects related to the different components of the water infrastructure. An important part of the professionals working in this area have a limited sectoral vision, identifying sanitation as water supply and sewage, when the problem is more complex and wide-ranging, where the components of flooding and urban drainage, solid waste and health cannot be ignored; Lack of management capacity: municipalities do not have the structure to properly plan and manage the different aspects of water in the urban environment (Tucci, 2004. p. 99).

In view of this problem, the process of consolidating the RMGSL must consider the Hydrographic Basins as planning and management units in accordance with Federal Law No. 9.433/1997, which considers that water resources, whether surface or underground, are public goods of common interest, over which everyone has the right to access and use to meet their various needs, and are therefore "[...] a preponderant issue for the quality of life of a people, of a nation" (OLIVEIRA, 2011).

With regard to the integrated management of water resources, the committees have emerged as a central figure. In the RMGSL, there are only the Mearim and Munim committees, both of which have yet to carry out their actions adequately because they do not have a structure that takes into account all the issues that arise in the BH area.

In this context, it is essential to create and strengthen basin committees and encourage effective participation by municipalities. The committees play a central role in the water resources management system, as they are political decision-making bodies with normative, deliberative and consultative powers in relation to the use, protection and recovery of water, involving public authorities, users and civil society.

As provided for in Law 9.433/97, it must be made up of the Union, the States and the Federal District, whose territories are located, even partially, in their respective areas of operation; the municipalities located, in whole or in part, in their area of operation; the water users in their area of operation and; civil entities for water resources with proven activity in the hydrographic basin, with the aim of enabling public management in the metropolitan area, where the solution of common problems requires joint policies and actions.

In view of the above, an integrated management of water resources must be implemented within the RMGSL with the consolidation of consortium operations, with the aim of guaranteeing regional development based on the peculiar characteristics of each municipality with a focus on public functions of common interest.

Cultural Heritage

The historical process of occupation of the northern coast of Maranhao, based on the export economy of agricultural products such as cotton and sugar, presents ancient settlement centers on the island of Maranhao and in the municipalities of the mainland, such as Rosàrio and Icatu - which originated from the old Arraial de Santa Maria de Guaxenduba founded by Jerônimo de Albuquerque - with remnants of the Forts of Santa Maria and Vera Cruz and other historical settlements, indicated and recognized in cartography and historical documents, dating back to the 17th, 18th and 19th centuries (Figure 20).

The municipalities that make up the RMGSL have a common history and share heritages and traditions that constitute a valuable cultural heritage, in part recognized and protected, with a Historic Centre included on the UNESCO World Heritage List (Sao Luis), a city recognized as a National Monument since 1948 (Alcântara), and folkloric manifestations recognized as National Intangible Heritage (Bumba-meu-boi and Tambor de Crioula) by the Federal Government (Chart 04).

Figure 20: Plan of Maranhao Island and its surroundings in 1615, by Joao Teixeyra Albernaz.

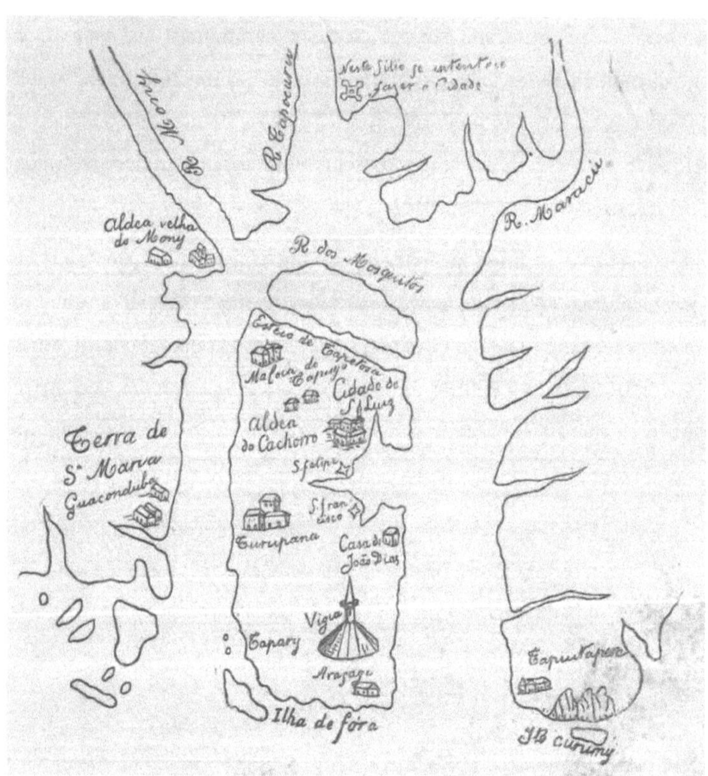

Source: MOURA, 1926.

The cultural manifestations present in the Metropolitan Region of Greater Sao Luis recognized as National Heritage are the Bumba-meu-boi and the Tambor de Crioula.

Bumba-meu-boi is a popular artistic manifestation based on the legend of Catirina, the pregnant wife of a humble cowboy who wanted to eat the tongue of the powerful landowner's favorite ox. A mixture of dance and theater, the bumba-meu-boi has variations related to the different regions of the state: the bois de matraca[10][11] , de zabumba[11] , de

---

10 In which the main percussion instruments are the matraca, two pieces of wood that are beaten against each other, and the rustic pandeiro, made of goatskin.

orquestra[12][13] , boi da baixada[1] 3 and de costa de mao[14] (IPHAN, 2011).

Table 04: Protected cultural heritage in the municipalities of Greater São Luis.

Municipality	Material Built Heritage	Intangible Heritage National
Alcantara	Ai'quitetônico e Urbanistico da Cidade de Alcântara (1973); Town Hall and Jail; House of the Holy Spirit; Fountain Ruins.	
Paço do Lumiar	Church and Square of Our Lady of Light.	Bumba-meu-boi; Tambor de Crioula.
Rosario	Calvary Fort; Railway station.	
Sao Luis	Ar'quitetônico e Urbanistico do Centro Historico de Sào Luis (1973); 45 Individual listed buildings of cultural interest.	

Source: SECID, 2016.

In the Greater Sao Luis area, the bois de matraca, which originated in the suburbs of the city of Sao Luis, and the bois de orquestra, which originated in the Munim region, where the municipalities of Axixà (Boi de Axixà), Morros (Boi de Morros) and Presidente Juscelino (Boi Brilho da Boa Hora and Boa Vista dos Pinhos) are located. The oxen of Bacabeira (Tradiçao de Peri) and Rosàrio are also famous.

The Tambor de Crioula do Maranhao is a form of Afro-Brazilian expression involving circular female dance, singing and drum percussion. It doesn't have a specific venue or calendar (Figure 21) and is practiced especially in praise of São Sebastiao and, although it is mostly concentrated in São Luis, there are groups active in other municipalities on the island of Maranhao and in the state. It is an important reference point for Afro-Maranhão cultural identity and resistance (RAMASSOTE, 2006).

Figure 21: Concentration of drum groups in Sao Luis, by locality.

---

11 With zabumbas, large drums of African origin punched by a mallet, complemented by nunchucks and tambourines. It is most present in the region of Guimaraes and its surroundings.
12 Accompanied by a band with wind and string instruments such as saxophones, banjos and clarinets.
13 It has a lighter, slower sound, marked by pandeiros and matracas, with Cazumba, animal and man, as the characteristic character.
14 The sound is marked by pandeiros played with the back of the hand, boxes and maracas. This accent originated in the Cururupu region.

Source: RAMASSOTE, 2006.

Bumba-meu-boi and Tambor de Crioula are manifestations that enrich and give rise to an identity peculiar to the June festivities in Maranhao, attracting a large number of locals and tourists to the festivals.

The June festivities in honor of Sao Joao and Sao Pedro (in the municipality of Sao Luis, Sao Marçal is also honored), traditionally held in the month of June, are a popular festival that is just as, if not more, significant than Carnival, and have great potential as a tourist attraction for all the municipalities of Greater Sao Luis.

The protection of heritage recognized by the Federal Government is carried out by the National Historical and Artistic Heritage Institute (IPHAN), an autarchy linked to the Ministry of Culture, with a Regional Superintendence (Parà, Maranhao and Piaui) based in Sao Luis.

In the state of Maranhao, the body responsible for the preservation and protection of the state's cultural heritage is the Secretary of State for Culture, through the Superintendence of Cultural Heritage and the Department of Historical, Artistic and Landscape Heritage of Maranhao (DPHAP), created by Decree No. 5.069 of March 11, 1973.

Of the 13 municipalities in the RMGSL, four have properties listed by the state or federal government. Of these, only Sao Luis has municipal legislation protecting cultural heritage and has a municipal body with jurisdiction over heritage, the Municipal Foundation for Historical Heritage (FUMPH), created in 2005. In addition to the Foundation, the municipality has a sub-prefecture with jurisdiction over the city's historic center.

Therefore, even in municipalities that have important heritage resources, the legal protection of material cultural heritage[15] and immaterial cultural heritage is not carried out by specific municipal legislation or by local bodies with jurisdiction over cultural heritage, such as secretariats, departments or councils, and it is the building departments that act in the general maintenance of buildings and public spaces in cities, treated indistinctly.

In addition to its cultural, material and immaterial heritage, the region also has important environmental resources that constitute a true collection of Brazilian environmental diversity, with beautiful sandy beaches and rivers that sustain many towns and communities on their banks.

The Rio Munim region, for example, is home to many quilombola communities that are fighting for ownership of their land and rural communities that preserve the values of rural workers in Maranhao. Expressions of the region's human and environmental diversity, these communities have their own particular ways of life, cuisine and religiosity and produce a rich and varied range of handicrafts (Chart 05).

Chart 05: Map of recognized cultural heritage in Greater São Luís.

MUNICIPALITY	QUILOMBOS	STATE ASSETS	FEDERAL PROPERTY
Alcântara	It has	It has	It has
Axixa	It has	None	None
Bacabeira	It has	None	None
Big waterfall	None	None	None
Icatu	It has	None	None
Hills	None	None	None
Paço do lumiar	None	It has	None
President Juscelino	None	None	None
Fox	None	None	None

---

15 Material cultural heritage here means the urban and architectural dimension of the territory, public spaces (streets, sidewalks, squares, squares, etc.), ruins, buildings, urban equipment and other elements that contribute to the development of the daily social practices of the city's inhabitants.

Rosario	It has	It has	It has
Santa Rita	It has	None	None
Sao José de Ribamar	It has	None	None
Sao Luis	None	It has	It has

Source: SECID, 2016.

Axixà is home to monuments such as Pedra do Tanque and Fazenda do Munim-Mirim. In Bacabeira, the crystal-clear springs of the village of Peri de Cima. In Morros, Presidente Juscelino and Cachoeira Grande, the charms of the Una and Munim rivers, with their waterfalls and lagoons.

Rosàrio is home to the Mother Church and the old Railway Station, as well as the ruins of the old Vera Cruz Fort. Santa Rita has a small spa and beautiful natural fields, as well as a 10,000-seat stadium that serves neighboring municipalities (SECID, 2016).

The municipalities of Morros and Icatu offer a gateway to the Lençôis Maranhenses region, with highly valued natural attractions and a rich handicraft activity based on the plant fibers of native species, as well as being part of the tourist route that integrates the coasts of Maranhâo, Piaui and Cearà, Piaui and Cearà, opening up a whole practically untouched region to tourism, including the beautiful beaches of Santa Marta, Papagaio and Prainha, as well as the Boqueirao waterfall, beauties that are currently isolated and almost inaccessible to tourists; the historic ruins of the Guaxenduba Fort, a remnant of the French occupation of Maranhao; and the Itatuaba River and the town of the same name, with its clean waters, wide riverbed and banks covered in riverine forest, all extremely valuable spots for ecological and adventure tourism (SECID, 2016).

In addition to the cultural and natural heritage assets of the municipalities that make up the metropolitan region, which are protected or listed by state and federal laws, there are also assets of tourist and cultural interest that are recognized as landmarks of their cities, even if they are not legally protected: these are small historic centers, beaches, rivers, squares and old buildings.

The assets of tourist and cultural interest distributed throughout the thirteen municipalities of Greater Sao Luis can become important resources for the socio-economic development of the region, through eco-tourism, nautical and sport fishing tourism, and adventure tourism, with events such as the automotive *rally*.

The greatest evidence that tangible and intangible cultural heritage is not yet considered an important resource in municipal public policies in the metropolitan region is the fact that there are no properties listed by municipal bodies, even in the city of Sao Luis. There is therefore no comprehensive policy for the conservation of buildings, uses and vitality of spaces as a way of preserving local memory and identity.

The challenge of urban preservation today lies mainly in taking into account the cultural, socio-political, economic and historical singularities and diversities of each city, while at the same time building the possibility of thinking about the management of the built heritage of the cities in the metropolitan region from a joint vision and planning.

TERRITORIAL STRATEGY

The new configuration of the Metropolitan Region of Greater Sao Luis brings together municipalities at different stages of economic and social development that make up a single territory. However, this composition is consolidated by the formatting and implementation of the new territorial strategy of the Maranhao State Government, with different perspectives for regional planning and management, where it will work in a sectorized way based on the characteristics and public functions of common interest.

Strategic Actions aim to:

- Integration of actions based on the implementation of Interfederative Governance;

- Articulation of public policies between federal entities linked to infrastructure for economic and social development, with the definition of structuring and priority projects;

- Consolidation of actions aimed at precarious and environmentally degraded urban areas with optimization of public services and facilities;

- Development of public policies of common interest to the municipalities of the Metropolitan Region;

- Execution of infrastructure aimed at linking central hubs;

- Technical support for plans, projects and access to tax incentives;

- Planning and management of works linking public and private services and facilities of metropolitan interest;

- Environmental preservation and tourism development.

The territorial strategies aim to strengthen the municipalities as federative entities, contributing to the consolidation of the RMGSL. These municipalities benefit from integrated planning of public functions of common interest, enabling them to plan, coordinate and manage policies and actions on a regional and local scale.

These strategies must be linked to investments associated with the provision of infrastructure, public services, job creation and professional training, tourism development, land use, subdivision and occupation, environmental protection, exploitation of water and mineral resources, agricultural production and food supply, affordable housing, and combating the causes of poverty and marginalization, thus promoting the integrated development of the region.

REFERENCES

BECKER, E. **Social-ecological systems as epistemic objects.** Institute for Social-Ecological Research (ISOE), Frankfurt/Main. Available at: http://www.isoe.de/ftp/publikationen/eb_socecsystem2010.pdf. Accessed on: Dec. 12, 2014.

BRAZIL, Federal Government. **Law 13.089 of January 15, 2015. Establishes the Metropolis Statute, amends Law No. 10.257, of July 10, 2001, and makes other provisions.** 2015

______. **Legislation. Law No. 11.107 of April 6, 2005.** Available at: <www.planalto.gov.br/legislaçao>. Accessed on: 28 Oct 2015

______. Ministry of the Environment. **Law No. 9.433, of January 8, 1997.** Institutes the National Water Resources Policy, creates the National Water Resources Management System, regulates item XIX of art. 21 of the Federal Constitution, and amends art. 1 of Law no. 8.001, of March 13, 1990, which amended Law no. 7.990, of December 228, 1989. Brasilia, DF, January 8, 1997.

______. **Research Report: Characterization and Comparative Analysis of Metropolitan Governance in Brazil**: institutional arrangements for metropolitan management (Component 1). Metropolitan Region of Greater Sao Luis. Rio de Janeiro: Institute of Applied Economic Research - IPEA, 2014.

CORREA, Roberto Lobato. **The Urban Network**. Sao Paulo: Editora Atica, 1989.

DINIZ, Joarez Soares. **The conditions and contradictions in the urban space of Sâo Luis (MA): peripheral traits.** Ciências Humanas em Revista - Sao Luis, V. 5, n.1, July 2007.

IBGE, BRAZILIAN INSTITUTE OF GEOGRAPHY AND STATISTICS. **2010 Demographic Census - Microdata**. Available at < http://www.ibge.gov.br/home/ > . Accessed on: September 2, 2013.

NATIONAL HISTORICAL AND ARTISTIC HERITAGE INSTITUTE - IPHAN. **Cultural Complex of the Bumba-meu-boi of Maranhao**: Dossier for registration as Cultural Heritage of Brazil. 210p. Sao Luis. IPHAN/MA. 2011.

IMESC, **Instituto Maranhense de Estudos Socioeconômicos e Cartogràficos. Gross Domestic Product (GDP).** Secretary of State for Planning and Budget. Sao Luis, 2012.

, Maranhão Institute of Socioeconomic and Cartographic Studies. **Preliminary Diagnosis of the MAIS IDH Action Plan.** Secretary of State for Planning and Budget. Sao Luis, 2015.

IPEA. **Social and urban infrastructure in Brazil: subsidies for a research agenda and public policy formulation**. Brasilia: Ipea, 2010.

. Atlas of Human Development in Metropolitan Regions. Brasilia: Ipea, 2013.

LIPIETZ, Alan. Capital and its space. Sao Paulo. Nobel, 1979.

KISHI, Sandra Akemi Shimada. **Integrated, participatory and decentralized water management**. Birigui, year 1, n. 3, p.40-41, 2011. Available at:<

http://midia.pgr.mpf.gov.br/4ccr/sitegtaguas/sitegtaguas_4/pdf/artigo1.pdf>. Accessed on: March 14, 2013.

MACROZEE, **Ecological-Economic Macrozoning of the State of Maranhao**.

Embrapa Satellite Monitoring, Sao Luis. 2013.

MARANHAO, **State Government. Complementary Law No. 174 of May 25, 2015**.

MASULLo, Yata Anderson Gonzaga; LoPES, José Antonio Viana. **The challenges of interfederative management in the face of social indicators in the metropolitan region of Greater Sao Luis - MA.** Rev. Tamoios, Sao Gonçalo (RJ), ano 12, n. 1, pág. 62-83. 2016.

MASULLo, Yata Anderson Gonzaga. **Evaluation of the spatial dynamics of dengue fever in relation to socio-environmental issues in the COHAB health district in the municipality of Sao Luis - MA.** Dissertation (Master's Degree) - State University of Maranhao, Postgraduate Program in Socio-Spatial and Regional Development, 2013.

MASULLO, Yata Anderson Gonzaga; MAGALHAES; Silvia Glacyane de Almeida. **Socioeconomic and Environmental Analysis of Conservation Units in the State of Maranhao.** Research Report. Foundation for Research Support and Scientific and Technological Development of Maranhao - FAPEMA. Sao Luis. 2015.

MOREIRA, Tiago Silva. **METROPOLITAN MANAGEMENT: the metropolitan region of Greater Sao Luis and the challenges of urban policies.** (Master's dissertation) Sao Luis: UEMA, 2013.

MOURA, Rosa. The reality of metropolitan management in Brazil. In: SINDICATO DOS

ENGENHEIROS DO ESTADO DO MARANHAO - SENGE. **Retrospective and Agenda of the Greater Sao Luis Metropolitan Forum**. Sao Luis: Raiz Comunicaçao Sustentâvel, 2012. p. 39.

MOURA, J. Abranches. The Island of S. Luiz. In: **Revista do Instituto de História e Geografia do Maranhao**. Year I. n°01. Sao Luis, 1926.

MUCHAILH, M.C. **Landscape analysis aimed at forming biodiversity corridors: A case study of the upper portion of the São Francisco Falso river basin, Paranâ**. Master's dissertation. Federal University of Paraná. Curitiba. 2007.

NUNES, G.M., Souza Filho, C.R.S., Vicente, L.E., Madruga, P.R.A; Watzlawick, L.F. **Geographic Information Systems applied to the implementation of ecological corridors in the Vacacai-Mirim River Sub-Basin (RS)**: In Anais do 12° Simpósio Brasileiro de Sensoriamento Remoto, Goiânia. 2005.

OBSERVATORY OF METROPOLISES. **Urban Territorial Units in Brazil Metropolitan Regions, Integrated Economic Development Regions and Urban Agglomerations in 2015.** Sao Paulo. 2015.

OLIVEIRA, Evagelina. **Health Networks and Regionalization**. In: BARCELLOS, Christovam. Geography and the context of health problems. Rio de Janeiro. ABRASCO. p. 223 - 230, 2008.

UN-HABITAT. **Habitat III Issue Papers**: urban and spatial design and planning. New York, May 31, 2015.

PRIMACK, R.B.; RODRIGUES, E. **Biologia da conservaçao**. Londrina. 2001.

ROLNIK, Raquel; SOMEKH, Nadia. **Governing the Metropolis: dilemmas of recentralization. Sâo Paulo em Perspectiva**, 2000.

RAMASSOTE, Rodrigo Martins (Coord.) **Os Tambores da Ilha**. Sao Luis: IPHAN, 2006. 125p.

SAQUET, M. **Approaches and conceptions of territory**. Sao Paulo: Expressao Popular, 2007.

SANCHO, Altair; DEUS, José Antonio Souza de. **Protected areas and urban environments**: New meanings and transformations associated with the phenomenon of extensive urbanization. Revista Sociedade e Natureza. Uberlândia. 27 (2): 223-238, May/August 2015.

SANTOS, L. C. A. dos. e LEAL, A. C. **Politica de recursos hidricos no estado do Maranhâo.** In: CUNHA, L; PASSOS, M. M. dos; and JACINTO, R.(Org). The new geographies of Portuguese-speaking countries: landscapes, territories, politics in Brazil and Portugal. Guarda: CEI, 2010.

SECID, SECRETARIAT OF STATE FOR CITIES AND URBAN DEVELOPMENT. **Report**: survey of environmental resources in the Greater Sao Luis Metropolitan Region. Sao Luis: SECID, 2016.

SEMA - SECRETARIAT OF STATE FOR THE ENVIRONMENT AND NATURAL RESOURCES OF MARANHAO. **Councillors approve the creation of the Mearim and Munim river basin committees.** 2013 a. Available at: < http://www.sema.ma.gov.br/paginas/view/paginas.aspx?id=1363>. Accessed on: March 12, 2013.

SOUZA, Barbara Cecilia Machado Fontes de. Regional development and metropolitan management: reflections on housing policy in the Aracajù metropolitan region. (Master's Degree in Regional Development and Management of Local Enterprises). Sao Cristóvao, Federal University of Sergipe, Sergipe: 2009.

SOUZA, Marcelo Lopes de. **Mudar a cidade: uma introduçâo critica ao planejamento e à gestão urbana.** Rio de Janeiro: Bertrand Brasil, 2010.

SERPA, Angelo. **Place and Centrality in a metropolitan context.** In: Carlos, Ana Fani Alessandrini. SOUZA, Marcelo Lopes de. SPOSITO, Encarnaçao Beltrao (Orgs). The production of urban space: agents and processes, scales and challenges. Contexto. Sao Paulo, 2012.

SILVA, Pollyanne Evangelista da. et. al. **Socio-environmental Vulnerability indices for the municipalities of Rio Grande do Norte.** 2000 census data. XIX National Meeting of Population Studies. Sao Pedro, SP. 2014.

UNION OF ENGINEERS OF THE STATE OF MARANHÃO - SENGE. **Retrospective and Agenda of the Greater Sâo Luis Metropolitan Forum.** Sao Luis: Raiz Comunicação Sustentâvel, 2012.

SOUZA, Marcelo Lopes de. **Mudar a cidade: uma introduçâo critica ao planejamento e à gestão urbana**. Rio de Janeiro: Bertrand Brasil, 2010.

SOUZA, Gustavo de Oliveira Coelho de. **CONSTRUCTION OF A SOCIAL AND ENVIRONMENTAL INDICATOR: the example of the Municipality of São Paulo.** Sao Paulo em Perspectiva, v. 20, n. 1, p. 61-79, Sao Paulo. 2006.

TABARELLI, M.; GASCON, C. **Lessons from fragmentation research: improving policies and management guidelines for biodiversity conservation.** Megadiversity Magazine 1. 2005.

TUCCI, C. E. M. (2004). **Integrated management of urban flooding in Brazil**. Rega/Global Water Partnership South America, v. 1, n. 1 (jan/jun, 2004). Santiago, GWP/South America

TUNDISI, José Galizia; MATSUMURA-TUNDISI, Takako. **Water Resources in the 21st Century.** Sao Paulo: Oficina de Textos, 2011.

UEMA - MARANHAO STATE UNIVERSITY. Geoenvironmental Center. **Hydrographic Basins:** Subsidies for Territorial Planning and Management. Sao Luis: UEMA, 2011.

ANNEXES

Annex 1

METROPOLITAN REGION WORKING group

OF GREATER SAO LUiS

Alan Linhares	Bacabeira City Hall
Adriana Aguiar Batista	Axixà Town Hall
Adriana Fabiola Martins Sousa de Jesus	Municipal Department of Articulation and Development Metropolitan - Sao Luis
Adriano Leonardo R. Lima	Communications Office - ASCOM/Gab. Ver. Rose Sales
Alfredo A. Costa Neto	Deputy Secretary for Urban Development -
Ana Paula dos Santos do Vale	SADU/SECID
Anderson Viana	Department of the Environment - Bacabal
Andréia Feitosa	Department of the Environment - Bacabal
Antonia Apoena Rejane da Silva Ribeiro	Department of Infrastructure - Paço do Lumiar Sao José de Ribamar City Hall
Apoena Ribeiro	Sao José de Ribamar City Hall
Balbuia Maia Rodrigues de Deus	Paço do Lumiar City Hall
Barbara Irene Wasinski Prado	State University of Maranhao - UEMA
Bruno Leonardo Silva Rodrigues	Attorney General's Office
Claudene do Socorro Campos	Department of the Environment and Natural Resources - Paço do
Claudio Eduardo de Castro	Lumiar
Crezus Ralph Larra Santos	State University of Maranhao - UEMA
Danielle Freitas	Alcântara City Hall
Danilo Almeida Castelo Branco	Communications Office - ASCOM/Gab. Ver. Rose Sales
Domingos Araken Santana da Cunha Junior	Municipal Department of Works and Public Services - Sao Luis Alcântara City Hall
Edinaldo Oliveira Moura	Municipal Secretariat of the Acceleration Program
Emanuel Pereira	Growth - Paço do Lumiar
Eny de Jesus M. Cardoso Sà	Rosàrio Town Hall
Èrica Garreto Ramos Barbosa	State Secretariat for Cities and Urban Development - SECID City, Research and Urban and Rural Planning Institute - Sao Luis
Evandro Tito Ferreira Soares	Municipal Department of Works and Public Services - Sao Luis
Fabiano Junqueira Ayres	Municipal Department of Works and Public Services - Sao José de
Fernando Dualibe Mendonça	Ribamar
Fredson Froz	Maranhao State Federation of Municipalities - FAMEM
George Henrique Machado Cunha	Department of the Environment - Sao José de Ribamar

Hugo Miranda Barbosa	Municipal Attorney General's Office - Rosàrio
Iellen Linhares	Bacabeira City Hall
Ilan Kelson de M. Castro	Rosàrio Town Hall
Irlahi Linhares	Maranhao State Federation of Municipalities - FAMEM
Jaqueline Aguiar da Silva	Rosàrio Town Hall
José Antonio Viana Lopes	Paço do Lumiar City Hall
José Carlos Muniz Lobato	Deputy Secretary for Metropolitan Affairs -
José de Ribamar Pinheiro Jûnior	SAAM/SECID
José Fortunato Filho	Morros Town Hall
José Marcelo do Espirito Santo	Bacabeira City Hall
José Raimundo Trindade	City, Research and Urban and Rural Planning Institute - Sao Luis
Josimar Sobrinho Oliveira	City, Research and Urban and Rural Planning Institute - Sao Luis
Júlio César Marques	National Union for Popular Housing
Lidiane Rosa	Paço do Lumiar City Hall
Liene Soares Pereira	Municipal Department of Articulation and Development
Lise Lucia Leite Coramx	Metropolitan - Sao Luis
Marcio Ribeiro Machado	Raposa City Council
Marcus Aurélio Borges Lima Maria	State Secretariat for the Environment and Water Resources -
Lais da Cunha Pereira	SEMA
Maria Madalena dos S. P. Xavier	Icatu Town Hall
	Municipal Department of Urban Planning and Housing - Sao Luis
	Municipal Attorney General's Office - Sao José de Ribamar
	Institute of Architects of Brazil - IAB and Council of
	Architecture and Urbanism - CAU-MA
	Sao José de Ribamar Environment Department
Maria Zilene N. da Silva Braga	East Maranhão Intermunicipal Development Consortium
Nonato Lima	Sao José de Ribamar City Council
Patricia Vieira Trinta	City, Research and Urban and Rural Planning Institute - Sao Luis
Raimundo Francisco Bogéa Junior	Bacabeira City Hall
Regina Celia Trindade	ISMAMSEL
Renata Coqueiro	Maranhao State Federation of Municipalities - FAMEM
Renata Cristina Azevedo Portela	Maranhao State Federation of Municipalities - FAMEM
Ricardo Filho	Municipal Department of Works and Public Services - Sao Luis
Ronald Abreu	Department of the Environment - Paço do Lumiar
Ronald Henrique Gomes Chaves	ISMAMSEL
Rose Sales	Sao Luis City Council

Rosilene Pereira	Raposa City Council
Solânea S. Dias Araùjo	Municipal Attorney General's Office - Paço do Lumiar
Suely Gonçalves da Conceiçâo	State Council of Cities - MA
Thiago Penha	Maranhao State Federation of Municipalities - FAMEM
Valerio Rabêlo	Municipal Department of Works and Public Services - Sao Luis
Valten Guimaraes	Santa Rita City Hall
Victor dos Santos Viegas	Maranhao State Federation of Municipalities - FAMEM
Viviane Gomes de Brito	Attorney General's Office - Sao Luis
Wagno Costa Lima	Alcântara City Hall
Yata Anderson Gonzaga Masullo	Deputy Secretariat for Metropolitan Affairs - SAAM/SECID

Annex 2

EXECUTIVE POWER

COMPLEMENTARY LAW NO. 174, OF MAY 25, 2015

Provides for the establishment and management of the Metropolitan Region of Greater Sao Luis and repeals State Complementary Laws No. 038 of January 12, 1998, No. 069 of December 23, 2003, No. 153 of April 10, 2013, No. 161 of December 3, 2013 and other provisions to the contrary.

The GOVERNOR OF THE STATE OF MARANHÃO,

I hereby inform all its inhabitants that the State Legislative Assembly has enacted and I hereby sanction the following Complementary Law:

CHAPTER I

THE INSTITUTION OF THE METROPOLITAN REGION

GREATER SÂO LUiS

Art. 1 The Metropolitan Region of Greater Sao Luis - RMGSL shall be governed by the rules set out in this Complementary Law.

Art. 2 The Greater Sao Luis Metropolitan Region includes the municipalities of Alcântara, Axixâ, Bacabeira, Cachoeira Grande, Icatu, Morros, Presidente Juscelino, Paço do Lumiar, Raposa, Rosàrio, Santa Rita, Sao José de Ribamar and Sao Luis.

Sole paragraph. The execution of public functions of common interest to the Region's member municipalities will be based on the RMGSL's Integrated Development Master Plan (PDDI).

Art. 3 The accession of new municipalities to the Greater Sao Luis Metropolitan Region must be based on prior technical studies, to be prepared by a public research institution with notorious knowledge and experience in regional and urban studies, which must be approved by the Metropolitan Collegiate, for subsequent submission to the Maranhao Legislative Assembly, considering the following criteria:

I - functional articulation between municipalities, with contiguity and/or discontinuity of the area of occupation (ports, airports, complex services, dormitory towns, research and innovation, major economic and infrastructure investments, landfills, water sources, etc.);

II - insertion in the region of influence of the city of Sao Luis, according to the Brazilian Institute of Geography and Statistics - IBGE (REGIC);

III - annual population growth rate above the state average (1.52% p.a. between 2000 and 2010);

IV - existence or need for public functions of common interest;

V - high tourist interest, environmental protection and cultural valorization;

VI - significant commuting of the population to work and/or study.

Sole paragraph. The municipalities that come to be constituted as a result of the dismemberment of a municipality belonging to the RMGSL are automatically integrated into the Metropolitan Region of Greater Sao Luis.

CHAPTER II

PUBLIC FUNCTIONS OF COMMON INTEREST

Art. 4 The Metropolitan Collegiate, based on the RMGSL Integrated Development Master Plan, will specify the public functions of common interest to the municipalities that make up the Greater Sao Luis Metropolitan Region, among the following functional fields:

I - the establishment of plans, programs and projects in the Integrated Development Master Plan for economic and social development;

II - basic sanitation, including water supply, sewage, drainage and solid waste services;

III - planning and land use;

IV - transport and the metropolitan road and waterway system;

V - environment and water resources;

VI - housing policy, land regularization and agricultural development;

VII - health, education and training of human resources;

VIII - tourism, culture, sport and leisure;

IX - public security and civil defense;

X - care and social assistance.

§ **Paragraph 1** Public functions of common interest are those whose execution requires inter-federative sharing of public agents, since they transcend the competence of municipalities because they affect an agglomerated space.

CHAPTER III

of the management of the metropolitan region of greater sao louis

Section I - General Provisions

Art. 5 The implementation of interfederative governance of the RMGSL, through collaboration, articulation and integration between the state and the municipalities of the metropolitan region, should fundamentally result in the following benefits:

I - optimizing the development potential and opportunities of the MR and disseminating their effects to the state as a whole;

II - reducing social and economic inequalities between municipalities and between social segments;

III - fair distribution of the benefits and burdens arising from the process of metropolization;

IV - consolidating metropolitan awareness and identity;

V - democratic management and social control.

Sole paragraph. In the interfederative governance of the RMGSL, the state and the municipalities that make up the metropolitan region will share responsibilities and actions for organizing, planning and executing public functions of common interest.

Art. 6 The RMGSL inter-federative governance structure is made up of:

I - Metropolitan Collegiate Body, with an executive and deliberative nature;

II - Metropolitan Participatory Conference and Council, with an advisory and deliberative nature;

III - Metropolitan Executive Agency, with a consultative nature and providing technical and operational support to the Collegiate and the Metropolitan Council;

IV - Metropolitan Development Fund.

Section II - Metropolitan Council

Art. 7 The Metropolitan Board is made up of:

I - State Governor;

II - Secretary of State for the Civil House;

III - Secretary of State for Cities and Urban Development;

IV - Secretary of State for Planning, Budget and Management;

V - Secretary of State for Political and Federal Affairs;

VI - Secretary of State for Infrastructure;

VII - Secretary of State for Health;

VIII - Secretary of State for Social Development;

IX - Secretary of State for the Environment and Natural Resources;

X - Secretary of State for Tourism;

XI - Secretary of State for Labor and Economic Solidarity;

XII - Secretary of State for Industry and Commerce;

XIII - Secretary of State for Education;

XIV - Mayor of the Municipality of Alcântara;

XV - Mayor of the Municipality of Axixà;

XVI - Mayor of the Municipality of Bacabeira;

XVII- Mayor of the Municipality of Cachoeira Grande;

XVIII - Mayor of the Municipality of Icatu;

XIX - Mayor of the Municipality of Morros;

XX   - Mayor of the Municipality of Paço do Lumiar;

XXI - Mayor of the Municipality of Presidente Juscelino;

XXII- Mayor of the Municipality of Raposa;

XXIII   -Mayor of the Municipality of Rosàrio;

XXIV   -Mayor of the Municipality of Santa Rita;

XXV- Mayor of the Municipality of Sao José de Ribamar;

XXVI   - Mayor of the Municipality of Sao Luis.

§   1° The resolutions of the Metropolitan Collegiate shall be approved by the favorable vote of at least 60% of its members.

§   **Paragraph 2 -** Each representative shall have an alternate to replace him/her in his/her absence or impediment, appointed by the head of the body represented.

Art. 8 The Metropolitan Board has the following competencies:

I - promote the preparation, monitoring and approval of the Integrated Development Master Plan and Sector Plans, as well as ratifying any revisions that may be necessary;

II - to forward, monitor and evaluate the execution of the Integrated Development Master Plan, as well as to approve any modifications that may be necessary for its correct implementation;

III - to specify, as a result of the dynamics of metropolization, new public functions of common interest within the RMGSL, as well as to promote their integration into the Integrated Development Master Plan;

IV - make compatible and deliberate on the application of resources from different sources, destined to fulfill public functions of common interest, in line with the RMGSL Integrated Development Master Plan;

V - establish the guidelines for the tariff policy for metropolitan services of common interest;

VI - promote the coordination of the municipalities of the metropolitan region with each other and with private organizations, federal and state bodies and entities, with the aim of planning and integrated management of public functions of common interest;

VII - to propose criteria for financial compensation to municipalities whose development is affected or which bear burdens arising from the execution of public functions of common interest;

VIII - forward, in good time, programs and projects from the Integrated Development Master Plan for inclusion/integration into the Multi-Year Plan, the Budget Guidelines Law and the Annual Budget Law of the State and the municipalities that make up the RMGSL;

IX - Set up Sectorial Chambers to analyze, debate and propose programs and projects related to public functions of common interest;

X - make public and give access to its work and decisions with the aim of making its activities transparent;

XI - promoting the drafting and approving of its Internal Regulations.

§ Paragraph 1 - The Metropolitan Collegiate shall make its decisions compatible with the guidelines set by the Union and the State for urban and regional development.

§ Paragraph 2 The Metropolitan Collegiate shall issue provisional instructions defining programs and projects for public functions of common interest until the Integrated Development Master Plan is approved.

Art. 9 Technical advice to the Metropolitan Collegiate will be provided by the Metropolitan Executive Agency of Greater Sao Luis.

Art. 10: The bodies or entities of the administrations of the state and the municipalities of the RMGSL shall not initiate or proceed with any request or negotiation for financial aid, loans, financing or the provision of services by public or private, national, foreign or international entities, related to investments in the Metropolitan Region of Greater Sao Luis, without the Metropolitan Collegiate certifying that the projects are in conformity with the Integrated Development Master Plan and other guidelines established for the RMGSL.

Section III - Participatory Council of the Metropolitan Region of Greater São Luis

Art. 11: The Participatory Council of the Metropolitan Region of Greater Sao Luis is composed of:

I - President of the Metropolitan Executive Agency of Greater Sao Luis;

II - 1 (one) representative of the Municipality of Alcântara;

III - 1 (one) representative of the Municipality of Axixà;

IV - 1 (one) representative of the Municipality of Bacabeira;

V - 1 (one) representative of the Municipality of Cachoeira Grande;

VI - 1 (one) representative of the Municipality of Icatu;

VII - 1 (one) representative of the Municipality of Morros;

VIII - 1 (one) representative of the Municipality of Paço do Lumiar;

IX - 1 (one) representative of the Municipality of Presidente Juscelino;

X - 1 (one) representative of the Municipality of Raposa;

XI - 1 (one) representative of the Municipality of Rosàrio;

XII - 1 (one) representative of the Municipality of Santa Rita;

XIII - 1 (one) representative of the Municipality of Sao José de Ribamar;

XIV - 1 (one) representative of the Municipality of Sao Luis;

XV - 2 (two) representatives of popular movements related to housing, sanitation or transportation issues;

XVI - 2 (two) representatives of workers' unions;

XVII- 2 (two) representatives of business organizations;

XVIII - 2 (two) representatives of professional councils;

XIX - 2 (two) representatives of university institutions;

XX - 3 (three) representatives of public utility companies (sanitation, lighting, transportation, etc.);

XXI - 4 (four) representatives of the City Councils of the RMGSL member municipalities.

Sole paragraph. The Participatory Council will have the technical and operational support of the Metropolitan Executive Agency for the organization and execution of its activities.

Art. 12: The purpose of the Participatory Council of the Metropolitan Region of Greater Sao Luis is to:

I - to draw up proposals for consideration by the other bodies of the Greater Sao Luis Metropolitan Region;

II   - consider relevant matters prior to deliberation by the Metropolitan Council;

III - propose the creation of working groups to analyze and debate specific issues;

IV - to convene public hearings and consultations on matters under its consideration.

§ **Paragraph 1 -** The recommendations of the Participatory Council of the Metropolitan Region of Greater Sao Luis shall be approved by the favorable vote of an absolute majority of its members.

§ **2)** Each representative shall have an alternate to replace him/her in his/her absence or impediment, to be appointed by the head of the body represented.

Art. 13: The representatives of the civil society segments will be chosen at a Metropolitan Conference regulated by the Metropolitan Collegiate, organized and coordinated by the Metropolitan Agency, for a term of 2 (two) years, which may be renewed at a new conference.

Sole paragraph. Candidates for membership of the Participatory Council of the Metropolitan Region of Greater Sao Luis must be recognized for their work in public functions of common interest to the Metropolitan Region, based and working in the same region and living in different municipalities.

Art. 14: The Participatory Council of the Metropolitan Region of Greater Sao Luis may set up Sectoral Technical Chambers, under the terms of the Internal Regulations.

Sole paragraph. The Sectorial Chambers are support units for the Participatory Council of the Metropolitan Region of Greater Sao Luis and the Metropolitan Executive Agency, created to provide technical support for sectorial specificities, in the debate, analysis and forwarding of proposals for projects and programs relating to public functions of common interest.

Section IV - Metropolitan Executive Agency

Art. 15: The Metropolitan Executive Agency, a state agency to be created by law, has the following powers:

I - liaise with the member municipalities of the RMGSL, with public and private, state, federal and international bodies and entities, with a view to combining efforts for integrated planning and the fulfillment of public functions of common interest, in the preparation of the Integrated Development Master Plan - PDDI;

II - consolidate information on the programs and projects of the Integrated Development Master Plan for inclusion in the Multi-Year Plan, the Budget Guidelines Law and the Annual Budget Law of the State and the Municipalities of the RMGSL;

III - to provide technical and organizational advice to the municipalities of the metropolitan region, accompanying the drafting and revision of master plans and land use, occupation and parceling laws, in order to make their contents compatible with the metropolitan interest expressed in the Integrated Development Master Plan - PDDI;

IV - promote diagnoses of the socio-economic reality at municipal and metropolitan level, in partnership with related state and municipal bodies and with the participation of civil society, with a view to subsidizing integrated planning;

V - prepare and maintain technical studies of regional interest and set up a database with up-to-date information necessary for planning and drawing up the programs and plans to be developed;

VI - promote the implementation of programs and projects established in the Integrated Development Master Plan - PDDI, as well as monitor and evaluate their execution, proposing to the Metropolitan Collegiate any necessary adjustments;

VII - liaise with national and international public and private institutions with a view to attracting investment or financing resources for the integrated development of the RMGSL;

VIII - Providing technical and administrative support to the Metropolitan Collegiate, as well as coordinating and coordinating the work of the Sectoral Chambers set up by it;

IX - manage the resources of the Metropolitan Fund, submitting the financial control instruments to the appraisal and deliberation of the Guidance and Supervision Committee made up of members of the Metropolitan Collegiate, the Participatory Council and the Metropolitan Executive Agency;

X - collect its own revenues or those delegated or transferred to it, including fines and tariffs for services rendered;

Sole paragraph. The executive management of the autarchy shall be exercised by 1 (one) President and 2 (two) Directors, who shall be assigned technical and administrative functions.

Section V -

Greater São Luis Metropolitan Region Development Fund

Art. 16: The RMGSL Development Fund is hereby created with the aim of financing structuring programs and projects, executing and operating public functions of common interest to the RMGSL, in accordance with the Integrated Development Master Plan - PDDI, aiming to:

I - improving the quality of life and the socio-economic and environmental development of the Region;

II - the improvement of municipal public services considered to be of metropolitan interest; and

III - the reduction of social inequalities within the metropolitan region.

Sole Paragraph. The area of application of the resources of the RMGSL Development Fund will cover the Municipalities that make up the Metropolitan Region.

Art. 17: Municipalities that are part of the RMGSL that do not comply with the provisions of this Complementary Law or do not comply with the resolutions of the Collegiate and Participatory Councils will not be able to receive resources from the Metropolitan Fund.

Art. 18: The RMGSL Development Fund's resources include:

I - budgetary resources from the state and the municipalities that make up the RMGSL, allocated to it by legal provision (PPA, LDO and LOA), through apportionment with a percentage of the Municipal Participation Fund (FPM) of each municipality that makes up the RMGSL, complemented by at least the same amount from the state government.

II - budget allocations or transfers from the Union intended for the execution of programs and projects under the guidance of the Integrated Development Master Plan;

III - national and international loans and other resources from aid and cooperation and intergovernmental agreements;

IV - funds from the financial return on loans made for investments in works, services and projects of metropolitan interest;

V - proceeds from credit operations contracted by the state or municipalities to finance works and services of common interest, and income from the investment of its resources in the financial market, among others;

VI - non-repayable grants made to the RMGSL Development Fund by national or international organizations, including non-governmental organizations;

VII - donations from public or private, national, international or multinational individuals or companies, and other occasional resources.

VIII 1° The percentage of the FPM to be contributed by each RMGSL municipality to the RMGSL Development Fund will be defined by the Metropolitan Collegiate.

IX 2. The RMGSL Development Fund may transfer resources to the State and Municipal Treasuries for the payment of amortization and charges of internal or external credit operations, destined for the Metropolitan Fund, which may be contracted by the State or the municipalities that make up the RMGSL, in accordance with the rules established by regulation.

X (3) Failure by the municipality to pay its contribution shall result in the state withholding the corresponding amount when making compulsory or voluntary transfers from the state to the municipality.

Art. 19: The RMGSL Development Fund is of an accounting nature, and its resources will be invested in the form of repayable financing and the release of resources without return, under specific conditions for each beneficiary, subject to the following requirements:

I - the program, project or investment to be financed or financially supported with resources from the Metropolitan Fund must be characterized as being of common interest in the metropolitan region;

II - the program, project or investment must be included in the Integrated Development Master Plan or, in its absence, in the metropolitan guidelines established for the metropolitan region;

III - the program, project or investment must be approved and prioritized by the Metropolitan Collegiate;

IV - the beneficiary of the funds must prove compliance with the legal requirements regarding public sector indebtedness, where applicable;

V - the program, project or investment must be related to:

a) financing of costs related to the preparation of a study or project linked to the Integrated Development Master Plan;

b) financing the implementation of a program or project included in the Integrated Development Master Plan;

c) research linked to the public function of common interest and the study of its impact on the quality of life in the metropolitan region;

d) financing structuring works of common interest.

Art. 20: The financing granted and the resources released by the RMGSL Development Fund are subject to the following general conditions: I - for repayable financing:

a) the value of the financing corresponds to a maximum of 80% (eighty percent) of the total value of the program, project or investment;

b) the beneficiary must provide the resources for the counterpart, which will be at least 20% (twenty percent) of the total value;

c) the grace period will be a maximum of thirty-six months, and may not exceed six months of the investment completion period;

d) the repayment period of the loan will be a maximum of ninety-six months and will begin in the month following the end of the grace period;

e) the financial charges relating to interest and monetary restatement will be established by regulation;

f) the form and frequency of repayments of principal and financial charges shall be defined in regulations to be drawn up by the decision-making body within one (1) year of the fund's creation;

g) the requirement for guarantees will comply with the relevant legal rules;

h) the penalties to be applied in cases of default or failure to comply with tax regulations will be established by regulation;

II - The release of resources without return will be decided by the Metropolitan Collegiate and the resources will be allocated to the execution of programs, projects or undertakings that are part of the Integrated Development Master Plan - PDDI;

III - release of resources as a form of financial counterpart assumed by the state in credit operations or in financial cooperation instruments whose purpose is to finance the execution of programs and projects that are part of the Integrated Development Master Plan - PDDI;

IV - the release of resources from the fund for structuring programs and projects is subject to a technical opinion on feasibility and purpose to be issued by the Metropolitan Executive Agency in accordance with the Integrated Development Master Plan - PDDI;

V - once the programs and projects have been certified, the municipalities will present the details of their projects, which will be monitored and evaluated at all stages, in order to issue a Technical File. The Ficha Tècnica characterizes the work plan to be developed, defining the amount and the physical-financial timetable, as well as other elements to make it viable.

Art. 21: The RMGSL Development Fund will be administered by the Metropolitan Executive Agency.

Sole Paragraph. The duties of the managing body and the financial agent shall be defined by regulation, subject to the provisions of the complementary law that provides for the establishment, management and extinction of state funds.

Art. 22: The Metropolitan Fund's budget and financial statements shall be drawn up in accordance with the provisions of Federal Law No. 4.320, of March 17, 1964, and Federal Complementary Law No. 101, of May 4, 2000.

CHAPTER IV

THE METROPOLITAN PLANNING SYSTEM

Art. 23: The Metropolitan Planning System is made up of the following plans:

I - instruments indicated in Article 4 of Law No. 10.257, of July 10, 2001, the Cities Statute;

II - instruments indicated in Article 9 of Law No. 13.089, of January 12, 2015, the Metropolis Statute;

III - Integrated Development Master Plan for the Greater Sao Luis Metropolitan Region;

IV - local sectoral plans;

V - Metropolitan Information System.

Art. 24: The Integrated Development Master Plan (PDDI) will contain the guidelines for metropolitan planning, including metropolitan sector plans and local sector plans, and will include at least

I - guidelines for public functions of common interest, including strategic projects and priority actions for investment;

II - the macro-zoning of the urban territorial unit;

III - guidelines on the coordination of municipalities in the subdivision, use and occupation of urban land;

IV - guidelines for the intersectoral coordination of public policies affecting the urban territorial unit;

V - the delimitation of areas with restrictions on urbanization aimed at protecting environmental or cultural heritage, as well as areas subject to special control due to the risk of natural disasters, if any;

VI - the system for monitoring and controlling its provisions.

§ **Paragraph 1** - The Integrated Development Master Plan (PDDI) corresponds to the Integrated Urban Development Plan (PDUI) established by Law No. 13.089, of January 12, 2015, the Metropolis Statute, and will cover all the municipalities that make up the Metropolitan Region and the metropolitan collar.

§ **Paragraph 2** In accordance with Law No. 13.089, of January 12, 2015, the Metropolis Statute, the process of drawing up the Integrated Development Master Plan and monitoring its application will be ensured:

I - the promotion of public hearings and debates with the participation of representatives of civil society and the population, in all the municipalities that make up the urban territorial unit;

II - publicizing the documents and information produced;

III - monitoring by the Public Prosecutor's Office.

§ Paragraph **3.** Once the procedures set out in the previous paragraph have been taken into account, the Metropolitan Executive Agency of Greater Sao Luis shall issue the following:

I - the metropolitan sectoral plan for housing and land regularization;

II - the metropolitan sectoral plan for urban mobility;

III   - the metropolitan sector plan for basic sanitation (water supply, sewage, drainage and solid waste);

IV   - other metropolitan sectoral plans, relating to public functions of common interest, under the terms of a decision by the Metropolitan Collegiate.

Art. 25: The Integrated Development Master Plan and the metropolitan sectoral plans may be drawn up by means of studies produced by the planning sectors of municipalities, university entities and foundations, or institutional and scientific development, or which have been drawn up or contracted by an organ or entity of the Direct or Indirect Administration of the Municipalities that make up the Metropolitan Region.

Sole paragraph. The Integrated Development Master Plan (PDDI) must be reviewed and updated at least every 10 (ten) years.

Art. 26: In addition to the metropolitan sectoral plan, the state or municipality that is part of the Metropolitan Region may draw up a local sectoral plan, which, unless otherwise stipulated in the metropolitan sectoral plan, will only become effective after it has been approved by the Metropolitan Collegiate.

Art. 27: The municipalities that make up the Metropolitan Region of Greater Sao Luis, the state and the regional agencies shall make their plans and programs compatible with the guidelines of the Integrated Development Master Plan or the metropolitan sector plan.

CHAPTER V

THE METROPOLITAN INFORMATION SYSTEM

Art. 28: The metropolitan information system will gather statistical, cartographic, environmental, geological and other relevant data for the planning, management and execution of public functions of common interest in the RMGSL.

Art. 29: The Metropolitan Executive Agency will be responsible for collecting, systematizing, disseminating, controlling and evaluating data from the Metropolitan Information System.

Art. 30: The RMGSL metropolitan information system shall:

I   - building a georeferenced cartographic base for the metropolitan region;

II   - model statistical data regarding the development of the region;

III - organize and centralize information that was isolated in federal, state and municipal government databases;

IV - define the indicators and indices for monitoring the development of the RMGSL;

V - to define and characterize the Metropolitan Region's Human Development Units (UDHs).

Sole paragraph. The Metropolitan Executive Agency shall make available, on an annual basis, updated information on the situation of the Human Development Units (UDHs) in the Metropolitan Region.

CHAPTER VI
FINAL AND TRANSITIONAL PROVISIONS

Art. 31: The secretariat and administrative support functions of the Metropolitan Collegiate and the Metropolitan Council will be carried out by the Metropolitan Executive Agency.

Art. 32: As long as the Metropolitan Collegiate does not decide otherwise, the regulation and supervision of public services of state or municipal ownership linked to public functions of common interest in the Metropolitan Region shall be carried out by state entities.

Art. 33 The State Executive Branch shall issue a decree regulating the functioning of the bodies mentioned in this Chapter and the creation and functioning of the Sectorial Chambers, and may also create other permanent or temporary bodies.

Art. 34: In order to meet the expenses arising from the application of this Complementary Law, the Executive Branch is authorized to:

I - open special credits up to the limit of R$ 2,000,000.00 (two million reais);

II - incorporate the budget classifications included in the credits authorized in item I of this article into the current budget, opening additional supplementary credits if necessary.

Based on the guidelines of Article 21 of Law No. 13.089, of January 12, 2015, the Metropolis Statute, the Integrated Development Master Plan must be drawn up and approved by state law within three (3) years of the RMGSL becoming effective.

Art. 36: The municipalities that make up the Metropolitan Region of Greater Sao Luis will make their master plans compatible with the Integrated Development Master Plan within 3 (three) years of the approval of this PDDI by state law.

Sole paragraph. The municipalities that make up the Metropolitan Region of Greater Sao Luis that do not have master plans must draw up their plans, making them compatible with the PDDI, respecting the deadline stipulated in the caput of this article.

Art. 37: Federal Law No. 13,089 of January 12, 2015, the Metropolis Statute, shall apply subsidiarily.

Art. 38: This Complementary Law fully repeals State Complementary Laws No. 038 of January 12, 1998, No. 069 of December 23, 2003, No. 153 of April 10, 2013, No. 161 of December 3, 2013, and any other provisions to the contrary or that are incompatible.

Art. 39: This Supplementary Law comes into force on the date of its publication.

I therefore order all authorities to whom the knowledge and execution of this Law belongs to comply with it and enforce it as fully as it is contained herein. The Honorable Secretary of the Civil House shall cause it to be published, printed and circulated.

PALACIO DO GOVERNO DO ESTADO DO MARANHAO, EM SAO LUiS, 25 DE MAY 2015, 194° DA INDEPENDÊNCIA E 127° DA REPÙBLICA.

FLAVIO DINO DE CASTRO E COSTA

Governor of the State of Maranhao

MARCELO TAVARES SILVA

Secretary of State for the Civil House

FLAVIA ALEXANDRINA COÊLHO ALMEIDA MOREIRA

Secretary of State for Cities and Urban Development

GOVERNMENT OF THE STATE OF MARANHÃO

SECRETARY OF STATE FOR CITIES AND DEVELOPMENT

URBAN - SECID

Flávio Dino

Governor

Flávia Alexandrina Coelho Almeida Moreira

Secretary of State for Cities and Urban Development

José Antonio Viana Lopes

Assistant Secretary for Metropolitan Affairs

Yata Anderson Masullo

Special Advisor

Arthur Lobão Carvalho Architect and Urban Planner **Harvey Spence L. Caldas** Web Designer **Luis Felipe Costa Cunha** Civil Engineer **Pedro Aurélio da Silva Carneiro** Civil Engineer **Priscilla Vieira S. Bueno Oliveira** Architect and Urban Planner **Tiago Silva Moreira** Geographer

Trainees

Artur Bezerra Environmental Engineering **Ingrid Amorim Ferreira** Architecture and Urban Planning

Larissa Fernanda dos Santos Silva Architecture and Urbanism **Leandro Leda Carvalho Lisboa Architecture** and Urbanism **Lohanne Caroline C. Domingues** Architecture and Urbanism **Otávio Bruno Silva da Silva** Civil Engineering **Paulo Henrique Correia Silva Sá Vale** Architecture and Urbanism **Stephani dos Santos Furtado** Architecture and Urbanism

Valéria Oliveira Nery Architecture and Urbanism **Yago Luis Cardoso Neto** Architecture and Urbanism